Electron Crystallography for Materials Research and Quantitative Characterization of Nanostructured Materials

MATERIALS RESEARCH SOCIETY
SYMPOSIUM PROCEEDINGS VOLUME 1184

Electron Crystallography for Materials Research and Quantitative Characterization of Nanostructured Materials

Symposia held April 14–16, 2009, San Francisco, California, U.S.A.

EDITORS:

Peter Moeck

Sven Hovmöller

Stavros Nicolopoulos

Sergei Rouvimov

Valeri Petkov

Milen Gateshki

Phil Fraundorf

SYMPOSIUM GG ORGANIZERS

Peter Moeck
Portland State University
Portland, Oregon, U.S.A.

Sven Hovmöller
Stockholm University
Stockholm, Sweden

Stavros Nicolopoulos
NanoMEGAS SPRL
Brussels, Belgium

Sergei Rouvimov
Portland State University
Portland, Oregon, U.S.A.

Phil Fraundorf
University of Missouri-St. Louis
St. Louis, Missouri, U.S.A.

SYMPOSIUM HH ORGANIZERS

Frank (Bud) Bridges
University of California-Santa Cruz
Santa Cruz, California, U.S.A.

David A. Keen
Rutherford Appleton Laboratory
Oxfordshire, United Kingdom

Igor Levin
National Institute of Standards and Technology
Gaithersburg, Maryland, U.S.A.

Thomas Proffen
Los Alamos National Laboratory
Los Alamos, New Mexico, U.S.A.

Materials Research Society
Warrendale, Pennsylvania

CAMBRIDGE UNIVERSITY PRESS
Cambridge, New York, Melbourne, Madrid, Cape Town,
Singapore, São Paulo, Delhi, Mexico City

Cambridge University Press
32 Avenue of the Americas, New York NY 10013-2473, USA

Published in the United States of America by Cambridge University Press, New York

www.cambridge.org
Information on this title: www.cambridge.org/9781107408203

Materials Research Society
506 Keystone Drive, Warrendale, PA 15086
http://www.mrs.org

First published 2009
First paperback edition 2012

Single article reprints from this publication are available through
University Microfilms Inc., 300 North Zeeb Road, Ann Arbor, MI 48106

CODEN: MRSPDH

ISBN 978-1-107-40820-3 Paperback

Cambridge University Press has no responsibility for the persistence or
accuracy of URLs for external or third-party internet websites referred to in
this publication, and does not guarantee that any content on such websites is,
or will remain, accurate or appropriate.

Financial support for Symposium HH was provided by the U.S. Department of Energy
(DOE) Office of Basic Energy Sciences under Grant Number DE-SC0001031.
The opinions, findings, conclusions, and recommendations expressed herein are
those of the author(s) and do not necessarily reflect the view of the U.S. DOE.

SYMPOSIUM GG

*Invited Paper

SYMPOSIUM HH

*Invited Paper

*Invited Paper

vii

PREFACE

SYMPOSIUM GG

Symposium GG, "Electron Crystallography for Materials Research," was held April 13–14 at the 2009 MRS Spring Meeting in San Francisco, California. It started off on Easter Monday with a very well attended half-day tutorial by Prof. Sven Hovmöller of Stockholm University and Dr. Stavros Nicolopoulos, President and Founder of the NanoMEGAS company. The tutorial material can be openly accessed at the websites of the "nanocrystal fingerprinting project" at Portland State University [1]. On the following day, we had eleven invited talks, two contributed talks, and four poster presentations. More than half of these contributions resulted in peer-reviewed papers that are included in this proceedings.

Electron crystallography and structural fingerprinting of nanocrystals by means of both precession electron diffraction and high-resolution transmission electron microscopy were comprehensively covered by contributions from ten countries and three continents. Some consensus was reached that precession electron diffraction and electron diffraction tomography are instrumental breakthroughs that will lead to *ab initio* determinations of unknowns (with high structural complexity!). Some of these unknowns may only exist as nanocrystals. An artist's rendering of this emerging consensus is given below. Financial support for our symposium was generously provided by Portland State University, the Oregon Nanoscience and Microtechnologies Institute, NanoMEGAS, Calidris, Crystal Impact, Fischione Instruments, Gatan, and Hummingbird Scientific.

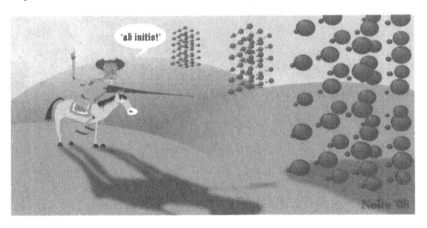

[1] nanocrystallography.research.pdx.edu

Peter Moeck
Sven Hovmöller
Stavros Nicolopoulos
Sergei Rouvimov
Phil Fraundorf

June 2009

PREFACE

SYMPOSIUM HH

Accurate knowledge of atomic arrangements in nanostructured materials is a key to understanding their physical properties. Unfortunately, traditional structure-solving approaches that assume long-range structural periodicity and rely upon Bragg reflections observed by x-ray/neutron diffraction methods fail on the local scale. Multiple experimental techniques exist for probing local atomic arrangements. Nonetheless, finding accurate comprehensive structural solutions for nanostructured materials still remains a formidable challenge because any one of the existing methods yields only a partial view of the local structure.

Symposium HH, "Quantitative Characterization of Nanostructured Materials," held April 14–16 at the 2009 MRS Spring Meeting in San Francisco, California, was motivated by the current lack of effective and robust measurement solutions to this truly interdisciplinary problem. The symposium's goal was to bring together experts in a wide variety of techniques used to probe the structure of nanomaterials, with an eye towards fostering collaborations that would combine techniques and thus provide better constraints on the nanostructure. About 85 presentations, including 12 invited speakers, 48 contributed talks, and 25 posters, highlighted the state of the art in nanostructure investigations using various x-ray/neutron scattering techniques (total scattering, small-angle scattering, COBRA), scanning probe microscopy (STM, AFM) for surface characterization, structural and compositional imaging in TEM, local electrode atom probe (LEAP), x-ray absorption fine structure (XAFS), neutron pair distribution function (PDF) analysis, NMR, Raman, and mass spectroscopy. Several presentations focused on the use of theoretical first principles calculations to interpret experimental measurements.

Financial support of this symposium was provided by MRS and the Department of Energy (DOE) Basic Energy Sciences. Peter Moeck, Valeri Petkov, Milen Gateshki, and Phil Fraundorf were so kind as to edit this part of the proceedings.

Frank (Bud) Bridges
David A. Keen
Igor Levin
Thomas Proffen

June 2009

MATERIALS RESEARCH SOCIETY SYMPOSIUM PROCEEDINGS

MATERIALS RESEARCH SOCIETY SYMPOSIUM PROCEEDINGS

Prior Materials Research Society Symposium Proceedings available by contacting Materials Research Society

SYMPOSIUM GG

Mater. Res. Soc. Symp. Proc. Vol. 1184 © 2009 Materials Research Society   1184-GG01-04

Quantitative Electron Diffraction for Crystal Structure Determination

Peter Oleynikov[1], Daniel Grüner[2,3], Daliang Zhang[1,2], Junliang Sun[1,2], Xiaodong Zou[1,2] and Sven Hovmöller[1]
[1]Structural Chemistry, Stockholm University, SE-106 91 Stockholm, Sweden
[2]Berzelii Centre EXSELENT on Porous Materials, Stockholm University, SE-106 91 Stockholm, Sweden
[3]Inorganic Chemistry, Stockholm University, SE-106 91 Stockholm, Sweden

ABSTRACT

We present a quantitative investigation of data quality using electron precession, compared to standard selected-area electron diffraction (SAED). Data can be collected on a CCD camera and automatically extracted by computer. The critical question of data quality is addressed – can electron diffraction data compete with X-ray diffraction data in terms of resolution, completeness and quality of intensities?

INTRODUCTION

X-ray crystallography has been an unprecedented method for determining atomic structures of inorganic crystals, organic molecules and proteins. If crystals of about 0.01 to 1 mm in all three directions can be obtained, then the structure can nearly always be solved to high accuracy, with atomic co-ordinates accurate to within about 0.1 Å for proteins, 0.01 Å for organic molecules and 0.002 Å for inorganic compounds. The main limitation of X-ray diffraction is to obtain sufficiently large single crystals. With the modern trend of nanotechnology, it is becoming increasingly interesting to be able to solve smaller and smaller crystals. Heavy developments are presently ongoing on synchrotrons, with the aim of making intense X-ray beams down to 1 μm in diameter. An alternative to single crystal X-ray diffraction is powder X-ray diffraction. In powder diffraction, the 3D pattern of diffraction spots is projected onto a 1D spectrum, creating big problems of overlapping diffraction peaks. This makes it very hard to solve complicated crystal structures by powder diffraction.

Electron diffraction can be obtained from crystals a million times smaller than the smallest that are feasible for X-ray diffraction, i.e. down to sizes of about 10x10x10 nm^3. Unfortunately, there are other problems with electron diffraction that has limited the usefulness of electron crystallography. Until now, it has been very difficult to collect complete and high-quality diffraction data by electron diffraction. The traditional method to collect electron diffraction data is selected-area electron diffraction, SAED. Because of the very short wavelength of electrons, 0.02-0.03 Å, compared to 0.5-2 Å for X-rays, the reflections that are simultaneously in diffracting condition lie almost on a flat plane. This makes it possible to record a geometrically undistorted diffraction pattern up to 1 Å resolution or higher in a single exposure in the electron microscope. The drawback of the very strong interaction of electrons with matter (about a million times stronger than X-rays) is multiple scattering. Unless the sample is extremely thin (<10 nm) a substantial fraction of the recorded electrons have scattered more than once, making the interpretation of the electron diffraction pattern very hard. However, when SAED patterns are taken from very thin edges, usually close to the edge of wedge-shaped crystals, diffraction data of acceptable quality can be obtained [1], although still not up to the quality of X-ray diffraction data.

The electron precession method [2] offers a possibility to acquire high-quality electron diffraction data, i.e. data that are less dynamical. Precession data are collected from one zone axis at a time. If we want to collect a complete 3D data set, it becomes very hard and time-consuming both with SAED and precession, since each zone axis has to be aligned manually in the transmission electron microscope and then processed one by one. Precession is easier to use than SAED, because the crystal orientation does not have to be quite as perfect for precession; the tolerance being < 0.1° for SAED but around 1° for precession.

For structures with one unit cell dimension about 4 Å, the crystal structure may be solved from a single projection along that short axis. Many metal oxides fall into this category. For crystals with the shortest unit cell dimension in the order of 6 to 10 Å, it is often necessary and sufficient to have the 3 orthogonal projections down the three crystallographic axes. Several recent papers demonstrate the power of electron crystallography for structure determination of sub-micron sized crystals of zeolites [3].

When all unit cell dimensions are > 10 Å, it becomes necessary to have data also from several projections [4]. For all types of structures it is necessary to have not only the intensities but also the crystallographic structure factor phases of sufficiently many reflections (typically roughly as many crystallographically unique reflections as there are unique atoms in the structure). These phases can be obtained from the Fourier transforms of high resolution transmission electron microscopy (HRTEM) images.

Ute Kolb has developed the electron diffraction tomography, where STEM data is collected from a range of tilt angles, using a computerized goniometer [5].

In this study, we compare the quality of different techniques of collecting electron diffraction data, including SAED and precession. The samples we used, $K_2O \cdot 7Nb_2O_5$ and an isotypic $Cs_2O \cdot 7Nb_2O_5$ have been studied by electron microscopy [6,7], while the thallium analogue was solved by X-ray crystallography [8].

EXPERIMENTAL DETAILS

Crystals of $K_2O \cdot 7Nb_2O_5$ were crushed, dispersed with ethanol and transferred on EM-grids. SAED patterns were collected on a JEOL 2000FX and a JEOL 2100LaB$_6$ TEM. Precession patterns were recorded on a JEOL 2000FX using the digital version of the SpinningStar from NanoMEGAS. Intensities of reflections were extracted from the selected-area electron diffraction and precession electron diffraction patterns by the program ELD [9] from Calidris. The structure was $K_2O \cdot 7Nb_2O_5$ was solved by direct methods using the program SIR97 [10]. The quality of the extracted data was evaluated using SHELX [11].

We used $K_2O \cdot 7Nb_2O_5$ for developing and testing new methods in electron crystallography. This niobium oxide has some advantages; it is very stable in the electron beam and has large unit cell dimensions ($a = b = 27.5$ Å, $c = 3.9$ Å). The high symmetry (space group $P4/mbm$) facilitates the evaluation of internal consistency of data; in the $hk0$ plane (shown in Figure 1) all the general reflections appear as 8 symmetry-related reflections, while the reflections on the axes $h00$ and diagonals $hh0$ appear four times. For 3D data, the general hkl reflections have 16 symmetry-related reflections.

4

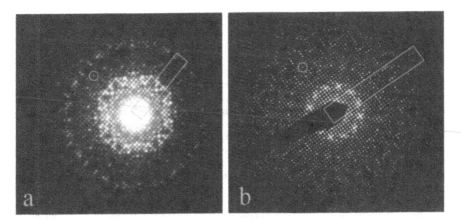

Figure 1. (a) Selected-area electron diffraction (SAED) and (b) electron precession pattern (0.8°) of $K_2O \cdot 7Nb_2O_5$. The reflection marked in both diffraction patterns is 18 0 0 which is at a resolution of 1.53 Å (=27.5 Å/18). A very slight (~0.2°) misalignment of the crystal is evident in (a) since the reflections go to higher resolution at the left part of the crystal. The precession pattern looks perfectly oriented. Although the SAED pattern is more severely affected by multiple scattering, the forbidden odd axial reflections (inside the rectangles) are clearly absent in the SAED (a) but present in the precession pattern (b). The local relative intensity distributions are quite similar, but the resolution is higher for precession, reaching 0.8 Å.

Figure 2. Kinematical electron diffraction pattern of $K_2O \cdot 7Nb_2O_5$ calculated using eMap [13]. The marked reflection is 18 0 0 at 1.53 Å resolution; it is the same one marked in Figures 1a and 1b.

5

The traditional way to collect electron diffraction patterns is selected-area electron diffraction (SAED) patterns, as shown in Figure 1a. An area of about 100 nm in diameter is selected. For most inorganic crystals, such a large area does not have a uniform thickness. This means that not only are there areas that are expected to be too thick for the simpler kinematical description to be valid, but the thickness will most certainly vary considerably over the selected area, such that a compensation for multiple scattering becomes hard or even futile. At 200 kV accelerating voltage, the wavelength of the electrons is so short (0.025 Å) that the Ewald sphere is nearly flat. This has the good effect that in a single exposure, an undistorted view of the entire $hk0$ plane can be seen up to and beyond 1.0 Å resolution. The negative side of this is that with so many reflections simultaneously excited, multiple diffraction may be maximized. At the higher resolution range, the effects of the curvature of the Ewald sphere are no longer negligible and as a result the high-resolution reflections look weaker than they really are, in comparison with the low-resolution reflections. This is evident in Figure 1a, where the reflections on and outside the ring of strong reflections are much weaker on the right part of the SAED pattern, because the Ewald sphere has been tilted ever so slightly over to the left. As a result the Ewald sphere passes over the high-resolution spots on the right hand side, while still passing through the reflections on the left. If high-quality quantitative structure factor data are to be collected, it becomes necessary to compensate SAED data both for the curvature of the Ewald sphere and the effects of even the slightest misalignments. To our knowledge, there are currently no such schemes being used.

The effects of multiple scattering are even worse than the effects of the curvature of the Ewald sphere, in that they differ from compound to compound, from crystal to crystal and from one area to another on the same crystal. In general, multiple scattering will result in the strong reflections redistributing their intensities over all other reflections. When the crystal is sufficiently thick, all reflections have virtually the same intensity, except for a general trend of falling off with scattering angle. This kind of electron diffraction pattern is often seen in the literature, probably helping to keep the pessimistic view on the usefulness of electron diffraction data as measures on the crystallographic structure factor amplitudes. However, it has long been known that it is possible to crush most inorganic crystals to such a fine powder that at least some of the crystals have areas near the edge that allow a decent electron diffraction pattern to be collected, i.e. one where the intensities are at least qualitatively related to the square of the structure factor amplitudes.

One way to evaluate the effect of multiple scattering is to look at the systematically forbidden reflections. Such reflections are found along screw axes and glide planes. The most common cases are such that all odd axial reflections are forbidden. This is also the case in the compound studied here, as seen in Figure 1a. If there is intensity in such forbidden reflections, it can only be because of multiple scattering. Unfortunately, the reverse is not true, i.e. one might think that the multiply scattered electrons can be neglected if the forbidden reflections are really absent. However, if the crystal is very close to perfectly aligned, as is the case for Figure 1a, the forbidden reflections will have zero intensity even for fairly thick crystals. The reason for this is the symmetry. The very same symmetry that causes some reflections to be forbidden, results in the effects of multiple scattering from either sides of the axis being identical in amplitude, but having phases exactly differing by 180°, such that they cancel out [13]. If the crystal is tilted slightly around that axis, then the reflections on one side will be more strongly excited than the ones on the opposite side and the multiply scattered electrons will no longer cancel out. The forbidden reflections will appear. This also explains the puzzling fact that sometimes the forbidden reflections are beautifully absent along one crystal axis, but appear quite clearly on the other axis – this is because the crystal is well aligned along one of the axes but tilted off the other axis.

Electron precession was invented by Roger Vincent and Paul Midgley [2] as a way to minimize the effects of multiple scattering. The electron beam is tilted on purpose, by an amount of up to about 3°. This will dramatically reduce the number of diffracted beams that are excited at the same time. By rotating the beam in a circular fashion, all the reflections around the optical axis are allowed to contribute to the final diffraction pattern, as seen in Figure 1b. One result is that the diffraction pattern will extend to higher resolution than an SAED pattern. Another effect is that every reflection is obtained as integrated over the entire spot, rather than the static cut through each reflection that occurs in SAED.

The effects on multiple scattering in precession are less trivial. In spite of the circulating misalignment, the innermost reflections will remain in scattering position all the time. Thus, they will be as much affected by multiple diffraction as those in an SAED pattern. Furthermore, since they are all the time diffracting, they will be very much stronger than reflections at higher resolution. This is evident by comparing the precession pattern in Figure 1b with the simulated kinematical electron diffraction pattern (Figure 2), where the first 9 diffraction orders (corresponding to 3.0 Å resolution) are very much stronger than those further out. For the systematically forbidden reflections, precession is not so favourable. The crystal is always tilted, so the cancellation of opposite contributions to forbidden reflections, described above for well-aligned SAED patterns, will not take place. Furthermore, the forbidden reflections are always along a line, which can be described as a systematic row of reflections. Great parts of the reflections along a crystallographic axis are simultaneously excited, together with the reflections immediately nearby on either side of the axis. This provides a possibility for multiple diffraction as is also evident in Figure 1b. The systematically forbidden reflections are as strong as the allowed ones all the way out to 18 0 0 (the reflection marked with a yellow circle). From then on and further out, the forbidden reflections have zero intensity (with an unhappy exception for 31 0 0). However, this should not be taken as a proof that multiple scattering is as serious in precession as in SAED. On the contrary, a practical demonstration of changing the precession angle from zero degrees (i.e. an SAED pattern is seen) to for example 3°, shows both how the diffraction pattern extends to higher resolution, and how equally strong reflections at zero precession angle suddenly sort out into strong and weak reflections at the higher precession angles.

We have attempted to evaluate the quality of precession data by quantifying the precession pattern in Figure 1b, using ELD [9]. The diffraction extends to beyond 34 diffraction orders, corresponding to 0.8 Å (27.5 Å/34), which is comparable to what is often reached in X-ray diffraction. There are 459 crystallographically unique reflections; only 1/8 of all the data seen in the precession pattern are independent due to the $4mm$ symmetry. The structure could be solved by direct methods from the precession electron diffraction data using the program SIR97 [10]. All the eight unique niobium atoms could be located from the potential map (Figure 3). Because of the mirror plane perpendicular to the 4-fold axis, all atoms have to lie exactly on 0 or ½ along z and thus the entire 3D structure can be solved from this single projection, making use of some very elementary chemical considerations. For example, along the z-axis, the niobium atoms all lie at z = 0. The oxygen atoms are between the Nb atoms and are either below or above the Nb atoms, i.e. at z = ½ or lie at z = 0 to form octahedra or pentagonal bipyramids around the niobium atoms.

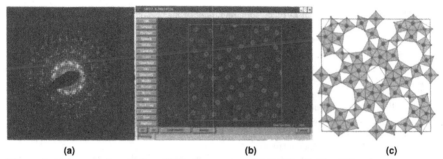

| (a) | (b) | (c) |

Figure 3. (a) a precession electron diffraction pattern of $K_2O \cdot 7Nb_2O_5$. The diffraction data extend to 0.76 Å resolution with a precession angle of 1.1°. The total observed number of unique reflections is doubled for precession compared to SAED. (b) The projected potential map obtained from the diffraction data by direct methods using the program SIR97 [10]. All the Nb positions (green circles) could be located. (c) The structure model of $K_2O \cdot 7Nb_2O_5$. The oxygen atoms lie between the Nb atoms.

The eight unique niobium atoms, two potassium and 20 oxygen atoms were refined against the intensity data extracted from Figure 1b. We refined individual isotropic temperature factors for each of the metal atoms, but kept a common temperature factor for all the oxygen atoms. The nice things about the refinement was that the metal atoms stay close to the positions determined by X-ray crystallography for the isotypic thallium compound [8] and that the crystallographic R-value (Sum $||F_{obs}| - |F_{calc}||/$ Sum$|F_{obs}|$) is below 30%. The bad news is that we could not get the R-value further down, not even to the level of 15% as was achieved for $Ti_{11}Se_4$ using SAED data [1].

DISCUSSION

In principle, electron diffraction can be used for solving crystal structures at atomic resolution, just as X-ray diffraction. Although there have been several examples of such studies, electron crystallography still demands too much of the operator. It is necessary to improve the technique, to allow automatic collection of complete 3D electron diffraction data to high resolution (1 Å or better for organic and inorganic crystals). With such data, it should be possible to solve unknown structures and refine the atomic co-ordinates to an accuracy similar to what we expect from X-ray diffraction, i.e. within 0.01 Å.

SAED data from inorganic crystals, collected at the edge of wedge-shaped crystals can be of sufficiently high quality for solving and refining the structure. The atomic co-ordinates may be in the order of 0.02 Å from their correct positions [1]. The crystallographic R-value is hard to get down below 15%, a value that is considered unacceptably high for X-ray crystallography. There may be several reasons for such high R-values; unreliable intensities due to multiple electron scattering in the crystals, imperfections in CCD detectors and in the computer programs used to quantify the diffraction intensities, insufficient corrections for geometrical factors (i.e. different diffraction spots have been exposed shorter or longer times) errors in atomic scattering factors for electrons and so on. As more and more groups are collecting more and more electron diffraction data and evaluate this data quantitatively, we believe that the ultimate goal of turning electron crystallography into a routine procedure for solving the atomic structures from sub-micron sized crystals is now within reach.

All the data has to be processed by a computer program providing accurate orientation of the start- and end-points of each exposure, all reflection indices *hkl* must be determined and the integrated intensities measured. Finally, all the data from one crystal must be merged and scaled into one single 3D data set. If radiation damage is not a problem, all the data can be collected on a single crystal. For radiation-sensitive crystals, including proteins, the crystals may only survive for a single frame of 1°, so the data collection and merging becomes much more cumbersome. Such work is being pursued [14] using the precession techniques.

CONCLUSIONS

The rapid development of new techniques for collecting electron diffraction data provides a hope that soon electron crystallography will become a method as mature as X-ray crystallography. For this it is necessary to develop hardware that allows automatic collection of full 3D electron diffraction data to at least 1 Å resolution and software that evaluates the data.

ACKNOWLEDGEMENTS

This research was carried out with financial help from the Swedish Research Council (VR) and the Swedish Governmental Agency for Innovation Systems (VINNOVA). We gratefully acknowledge the Knut and Alice Wallenberg Foundation for financial support for the purchase of some of the equipment employed in this work.

REFERENCES

1. T. Weirich, R. Ramlau, A. Simon, X.D. Zou and S. Hovmöller, Nature 382, 144-146 (1996).
2. R. Vincent and P. A. Midgley, Ultramicroscopy 53, 271 (1994).
3. C. Baerlocher, F. Gramm, L. Massüger, L.B. McCusker, Z.B. He, S. Hovmöller and X.D. Zou, Science 315, 1113-1116 (2007).
4. X.D. Zou, Z.M. Mo, S. Hovmöller, X.Z. Li and K. Kuo, Acta Cryst. A59, 526-539 (2003).
5. U. Kolb, T. Gorelik, C. Hübel, M.T. Otten and D. Hubert, Ultramicroscopy 107, 507-513 (2007).
6. J.J. Hu, F.H. Li and H.F. Fan. Ultramicroscopy 41, 387–397 (1992).
7. D.N. Wang, S. Hovmöller, L. Kihlborg and M. Sundberg, Ultramicroscopy 25, 303–316 (1988).
8. V. Bhide, M.Gasparin. Acta Cryst. B35, 1318–1321 (1979).
9. X.D. Zou, Y. Sukharev and S. Hovmöller, Ultramicroscopy 49, 147-158 (1993).
10. A. Altomare, M.C. Burla, M. Camalli, G. Cascarano, C. Giacovazzo, A. Guagliardi, A.G.G. Moliterni, G. Polidori and R. Spagna. J. Appl. Cryst. 32, 115-118 (1999).
11. G.M. Sheldrick, Acta Crystallogr. A64, 112-122 (2008).
12. J. Gjønnes and A.F. Moodie, Acta Crystallogr. 19, 65-67 (1965).
13. P. Oleynikov, Analitex, Sweden, http://www.analitex.com/.
14. D.G. Georgieva, L. Jiang, H.W. Zandbergen and J.P. Abrahams, Acta Crystallogr. D64, (2009) (in press).

Mater. Res. Soc. Symp. Proc. Vol. 1184 © 2009 Materials Research Society   1184-GG01-05

Automated Diffraction Tomography Combined With Electron Precession: A New Tool for *Ab Initio* Nanostructure Analysis

Ute Kolb*, Tatiana Gorelik, Enrico Mugnaioli

Institut für Physikalische Chemie, Johannes Gutenberg-Universität Mainz, Welderweg 11, D-55099 Mainz, Germany

ABSTRACT

Three-dimensional electron diffraction data was collected with our recently developed module for automated diffraction tomography and used to solve inorganic as well as organic crystal structures *ab initio*. The diffraction data, which covers nearly the full relevant reciprocal space, was collected in the standard nano electron diffraction mode as well as in combination with the precession technique and was subsequently processed with a newly developed automated diffraction analysis and processing software package. Non-precessed data turned out to be sufficient for *ab initio* structure solution by direct methods for simple crystal structures only, while precessed data allowed structure solution and refinement in all of the studied cases.

INTRODUCTION

The rapidly developing nanotechnology urgently needs analytical tools to characterize nano-volumes. The crystalline structure of a material is a principal key for understanding its properties and therefore is the most desired piece of information. Well developed methods for structure analysis by X-ray single crystal diffraction are established and routinely used in many laboratories. Single crystal X-ray analysis requires crystals with a size of at least about 1 mm^3. Powder X-ray diffraction can access significantly smaller crystals but indexing and subsequent structure solution is often problematic due to peak overlap, the presence of additional phases, and a preferred orientation. Furthermore, the problem of peak overlap is particularly enhanced due to crystal-size driven peak broadening for nanocrystalline materials.

High resolution transmission electron microscopy (HRTEM) is traditionally used for nanostructural investigations. Extrapolating three-dimensional (3D) structural information from images requires special tomographic techniques, which are typically not optimized for electron beam sensitive materials, such as inorganic complex structures (i.e. zeolites) and organic crystals, which cannot sustain the high electron dose needed to collect the data from nano-volumes. While strong efforts were dedicated in recent years for the construction of aberration correctors in order to achieve sub-Ångstrom resolution in imaging [1], the usage of the equivalent information in reciprocal space, already providing such a resolution and easily available in any transmission electron microscope (TEM), is not so well developed.

Electron diffraction can probe volumes down to 20-30 nm in diameter, delivering 3D sub-Ångstrom structural information with good signal-to-noise ratio. Nevertheless, there are only a few software packages using low index zone axis patterns [2] and no hardware or formalism for dealing with patterns that were recorded though a tilt around an arbitrary axis available so far. Underdeveloped diffraction instrumentation and the absence of processing routines are major reasons for the electron diffraction technique to be inferior to modern X-ray diffraction techniques.

Traditionally, after a suitable crystal has been selected in the TEM imaging mode, a diffraction pattern can be recorded by switching into the diffraction mode. After the crystal has been oriented with a low index axis along the goniometer axis, a series of prominent diffraction patterns (typically of low index crystallographic zones) can be recorded by tilting it around this axis. Since the tilt steps between the single crystal patterns are known, a 3D network of reflections can be obtained, from which the lattice parameters of the structure can be calculated [3].

While the unit cell determination using electron diffraction is commonly accepted as a reliable technique that dates back to the 1960s [4], the employment of electron diffraction intensities for structure determinations is still controversial. Electrons, interacting with matter some orders of magnitude stronger than X-rays, tend to scatter several times while traversing through a specimen. Multiple diffraction events modify the final intensities of the reflections, so that such data cannot straightforwardly be used for structure solution anymore. These events are denoted as dynamical effects and usually are stronger in low index diffraction patterns, which are typically used during diffraction data collection via manual tilt series.

Recently, the precession electron diffraction (PED) technique was developed as an approach to reduce dynamical effects in electron diffraction patterns [5]. This method is based on the precession of the incident primary electron beam which is inclined away from the optical axis of the TEM. A diffraction pattern recorded in this mode is the sum of patterns produced by the precessing beam sequentially. The intensities of the reflections are integrated throughout the (reciprocal) volume covered by the precessing Ewald sphere and show reduced dynamic diffraction effects.

Although manually collected electron diffraction tilt series are rather poor and suffer from dynamical diffraction effects, some *ab initio* structure solutions were successfully undertaken on the basis of such data [6]. Each of the structure solutions published is a very impressive investigation, but none of the solution paths proposed so far can be used routinely and provides reproducible results for different classes of materials.

AUTOMATED DIFFRACTION TOMOGRAPHY

Recently an automated (hardware) module for electron diffraction data collection and processing (software) was developed in our research group ([7], [8]). Since the module is based on a collection of diffraction patterns at different tilt angles during specimen tilt, similar to how it is done in real space tomography, the method is called automated diffraction tomography (ADT).

ADT principle

The principle of the method is sampling the reciprocal space in small steps without any prior information on the orientation of the crystal. The only essential requirement is that the data is collected from the same crystal. In such a way high index crystallographic zones are typically recorded through a tilt around an arbitrary axis. As the information about the angular relationship between the diffraction patterns is available, the three-dimensional reciprocal volume can be reconstructed. Figure 1 shows a schematic sketch of the reciprocal space of a crystal during the ADT data collection process.

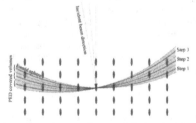

Fig. 1: Schematic sketch of the reciprocal space of a crystal during the ADT data acquisition process. The solid black lines correspond to the Ewald sphere positions for ADT pattern collection: the related diffraction patterns include the black segments. The precession mode integrates the data also between the tilting steps; the dotted lines stand for the borders of the PED integrated volumes at the optimal precession angle.

For simplicity, the reciprocal lattice is sketched here stationary, while the Ewald sphere, shown partly for three positions in Figure 1, is considered as moving with respect to it for each tilt step. Alternatively, one may consider the incident precessing primary electron beam as always creating the same hollow cone of illumination and the reciprocal crystal lattice rotating when the crystal is tilted in real space. As the reflections have a physical size, the Ewald sphere cuts them at certain heights, so that the corresponding diffraction patterns include the black segments drawn through the reflections. The data collected in such a way may suffer from gaps in the reciprocal space reconstruction between the tilt steps. PED helps to overcome this problem. Ideally, the precession angle should be equivalent to one half of the tilt step, so the intensity information between the tilt positions will be preserved within the PED patterns.

ADT acquisition

Before the data acquisition can be started, a series of instrumental calibrations has to be performed. The calibration procedure is described in detail in ref. [7]. The sample is imaged in the scanning transmission electron microscopy (STEM) mode. The diffraction is created employing the nano electron diffraction (NED) mode, i.e. a small condenser aperture is used in order to achieve a quasi-parallel electron beam of 70-30 nm in diameter. Usually both non-precessed NED and PED patterns are acquired.

First the sample stage is driven to the starting tilt angle. Then an appropriate crystal is selected, the first diffraction pattern is recorded, and the stage is tilted further. To ensure that after a tilt step the diffraction data is still collected from the same crystal, a crystal tracking procedure is implemented. The steps diffraction acquisition – tilting – tracking are then repeated in a loop until the desired part of the reciprocal space is scanned. Typically, the data is collected with a tilt step of 1°. Larger steps do not ensure a reliable sampling of the reciprocal space, leaving "holes" in the data; smaller steps are not necessary in practice for standard tasks. As specifications of the goniometer state, reliable tilt steps down to 0.1° are possible. Thus the true limitation in the angular accuracy is crystal bending, which for organic crystals can achieve up to 1° declination.

ADT processing

The collected ADT data set is processed by our newly developed automated data processing and analysis package (ADAP), already reported in ref. [8]. Subsequent to background correction and centering of the diffraction patterns with respect to the primary beam, an important block of processing routines is aimed at defining the correct geometry of the diffraction experiment. Especially a misaligned tilt axis leads to distortion of the complete diffraction volume, and can hamper unit cell determination, reflection indexing, and intensity integration. An experimental procedure is implemented in the acquisition module for a rough determination of the tilt axis position, which is often not precise enough. In order to refine the tilt axis position, the difference vector space is calculated for the collected set of diffraction spots. The difference vectors are sorted according to their direction in 3D space and plotted onto a sphere. Maxima on the sphere's surface represent directions to which most of the difference vectors are pointing.

Figure 2 (top) shows such a sphere calculated for a tilt series of orthorhombic $BaSO_4$ (tilt range of $\pm 60°$, tilt steps of 1°). The orientation map is shown on the logarithmic scale. One can see the basic vector directions of the reciprocal lattice (marked in the figure by black dashed arrows). The width of lines on the surface of the sphere is defined by a sum of geometrical imperfections of the data originating from the so called V-declination of the sample holder [9] and crystal bending.

For clarity, the sphere is projected onto a plane given by the angular coordinates θ (0° to 180°) and φ (0° to 360°). The bottom of the figure shows projected orientation maps that were calculated using the proper tilt axis position (left) and a tilt axis that was 5° off (right). The left map appears sharper, while in the right map the contrast is blurred and noisy. For the true determination of the correct tilt axis position, the traditional "sharpness" criterion is used, i.e. the variance of the map. The procedure seems to be robust and typically finds the correct position *ab initio* for all data sets (except for the most problematic ones), within an accuracy of up to 0.1°. An orientation map also delivers information about the cell parameters, main axes orientations and the part of reciprocal space that is sampled. Once the proper geometry of the diffraction experiment is determined, the 3D reconstruction of reciprocal space can be calculated and the unit cell parameters can be defined. The cell parameter determination is performed in difference vector space since difference vectors represent an autocorrelation of the diffraction spot positions, and strongly enhance short reciprocal distances. The three shortest non-coplanar vectors correspond to the three real space lattice vectors. In contrast to single crystal X-ray diffraction data, where the positions of difference vectors are defined precisely, the difference vector space of the ADT data consists of agglomerates of points, so devoted routines have to be used to accurately calculate the lattice parameters using the centers of the clusters [8]. Table 1 gives determined high precision unit cell parameters for a wide variety of materials, some of which are highly beam sensitive with unit cells volumes of up to 4500 Å³.

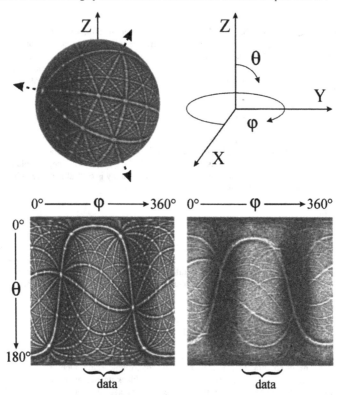

Fig. 2: Orientation histogram of the difference vector space for a $BaSO_4$ tilt series (tilt range of ±60°, tilt step 1°). Top: View of the orientation sphere and the coordinate system used. Bottom: Projections of this sphere onto a plane plotted in angular θ and φ coordinates; left side: using the proper tilt axis, right side: using a 5° misaligned tilt axis. The orientation maps show bright and dark vertical bands. Bright regions correspond to the angular regions in which the data was effectively acquired. Darker bands represent regions where no experimental data was collected, so the information presented in these regions is due to the difference vectors.

Table 1: Lattice parameters determined by ADT for different inorganic (upper part) and (metal-)organic (lower part) materials. The expected values are shown in brackets. For centered space groups, the primitive cell is reported.

	a (Å)	b (Å)	c (Å)	α (°)	β (°)	γ (°)	V (Å3)
Inorganics							
$CaCO_3$ (calcite) (R-3c) [10]	4.96 (4.99)	4.96 (4.99)	6.41 (6.38)	67.2 (66.98)	66.91 (66.98)	59.32 (60)	121.23 (122.60)
$BaSO_4$ (Pnma) [11]	8.89 (8.88)	5.51 (5.46)	7.17 (7.15)	90.1 (90)	89.6 (90)	90.5 (90)	351.21 (346.67)
Zn_1Sb_1 (Pbca) [12]	6.46 (6.54)	8.11 (8.06)	8.43 (8.31)	89.3 (90)	89.9 (90)	89.4 (90)	441.65 (438.04)
Charoite ($P2_1/m$) [13]	32.08 (32.30)	19.55 (19.65)	7.16 (7.20)	91.2 (90)	97.9 (96.3)	90.0 (90)	4446.88 (4542.21)
Organics							
P.Y. 213 (P-1) [14]	7.00 (6.90)	11.60 (11.83)	13.80 (14.06)	98.5 (98.2)	101.3 (99.0)	93.7 (92.5)	1044.20 (1091.83)
NLO ($Pca2_1$) [15]	28.15 (28.47)	5.15 (5.07)	11.09 (11.00)	90.2 (90)	89.8 (90)	90.1 (90)	1607.74 (1587.77)
NS3 ($P2_1/c$) [16]	14.72 (14.63)	9.98 (9.89)	12.59 (12.72)	89.6 (90)	107.4 (107.6)	89.7 (90)	1764.91 (1754.31)
Basolite (Fm-3m) [17]	18.60 (18.63)	18.58 (18.63)	18.71 (18.63)	60.84 (60)	61.48 (60)	60.29 (60)	4660.81 (4568.64)

Applying the determined unit cell vectors to a data set, the reflections can be indexed, and the intensity of the diffraction spots can be extracted. The intensities are integrated directly on the diffraction patterns. No correction factors are applied at the moment.

A full reconstruction of the 3D reciprocal volume gives valuable additional information about the structure. Although for conventional electron diffraction the kinematic extinction conditions are typically violated by dynamical effects, in 3D visualized ADT data extinctions both due to centering (integral) and glide planes (zonal) can be easily recognized. Figure 3a shows a view of the reconstructed 3D reciprocal space of trigonal $CaCO_3$ (space group R-3c).

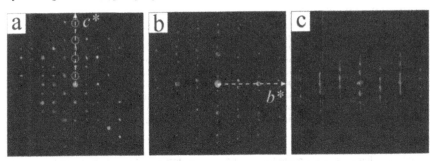

Fig. 3: Projections of fully reconstructed 3D reciprocal space: a) $CaCO_3$, showing extinctions due to a c-glide plane; b) $(Na_2O)_xTiO_2$ viewed along the c* axis: the C-centering of the structure is evident; c) $(Na_2O)_xTiO_2$ viewed along the a* axis: elongated reflections reveal one-dimensional disorder along c*.

One can clearly see missing rows of reflections along the c* axis, which correspond to c-glide plane in the CaCO$_3$ structure. Figure 3b presents the projection along the c* axis of the reciprocal volume of the monoclinic (Na$_2$O)$_x$TiO$_2$ compound [18]. In this projection a "chess-like" pattern of reflections results due to the C-centering of the lattice. Along the *-c* axis, the same structure is disordered, and the respective streaks can clearly be seen when viewed in the orthogonal direction (Figure 3c).

EXPERIMENTAL DETAILS

Fine powders of the materials were dispersed in ethanol, sonified and sprayed on a carbon coated copper grid with a UIS250v Hielscher sonifier [19] equipped with a caved tip for holding specimen dispersions. The transmission electron microscopy (TEM) analysis was carried out with a Tecnai F30 S-TWIN equipped with a field-emission gun and working at 300 kV. The diffraction data acquisition was performed with a GATAN single-tilt sample holder and a FISCHIONE tomography holder, using our ADT acquisition module developed for FEI Tecnai F30 TEM. The module works in the μ-probe STEM mode, and includes routines which allow to track the crystal after each tilt step and to acquire NED patterns sequentially. The ADT data has been collected with and without PED in steps of 1° in a range of ±30° and ±60°. For beam sensitive samples, STEM images and diffraction patterns were collected with a "mild illumination" setting (i.e. gun lens 8 and spot size 6 for our TEM) resulting in the electron dose rate of 10-15 e/Å^{2}s, and using a 10 μm C2 aperture. NED was performed employing a 50 nm beam producing semi-parallel illumination of the sample.

STEM images were collected by a FISCHIONE high angular annular dark field detector (HAADF). NED patterns were taken with a CCD camera (14-bit GATAN 794MSC) and acquired by the Gatan DigitalMicrograph software. All collected electron diffraction tilt series were saved in the MRC file format [20].

PED was performed using the "Spinning Star" unit developed by the NanoMEGAS company [5]. The precession angle was set to 1° and the precession frequency was 100 Hz.

The self-developed ADAP package [7], [8], programmed in Matlab, was used for data processing, including the 3D diffraction volume reconstruction and automated cell parameter determination procedure. For the visualization of 3D pictures we used the UCSF Chimera software [21], [22].

The "ab initio" structure solution was performed by SIR2008, which is included in the package Il Milione [23]. The structure refinement was done in SHELX97 [24].

PROCEDURES AND RESULTS

For a number of compounds an *ab initio* structure solution was carried out using direct methods implemented in SIR2008, which is included in the IL MILIONE package [23]. A fully kinematical approach was used (i.e. intensities proportional to F_{hkl}^2), and no correction was applied to the experimental data. Some case studies of crystal structures, of which four were known and two unknown (i.e. Zn$_8$Sb$_7$ and Charoite), are presented below and summarized in Table 2 (an the end of this section). The data resolution was between 0.8 to 1.1 Å, as a result of both the selected diffraction camera length (i.e. 380 or 560 cm) and the size of the employed CCD camera. Some of the crystallographic concepts that are used to quantify a structural solution in this paper are defined in the appendix.

Zn$_1$Sb$_1$

Zn$_1$Sb$_1$ was found as a side product phase during the synthesis of Zn$_8$Sb$_7$ (whose structure is also discussed below). The crystals were very small with a size of down to 20 nm and highly agglomerated as well as intermixed with the prominent crystal phase. Despite the strong aggregation, it was possible to collect ADT data sets from single nanocrystals in order to recognize the two phases by 3D reciprocal volume visualization and the subsequent unambiguous cell parameters determination. Since the Zn$_1$Sb$_1$ structure is relatively simple due to a high symmetry with only two strong "scatterers" in the asymmetric unit, a rather narrow tilt range of pure NED data was sufficient to find all atom positions correctly. The ADT data was collected within a total range of 60° and 388 reflections were extracted. After applying the orthorhombic symmetry, 106 independent reflections were left over. The space group Pbca (figure of merit: 0.152) was automatically recognized by the SIR2008 routines. (The next likely suggested space group: Pnca had a figure of merit of 0.035).

The structure solution was possible in one step with a residual value of 27.10 %. Four peaks of the electrostatic potential were identified by the program. The most intense peak with a height of 1596 was assigned to the antimony atom. The second peak with the height of 910 was assigned to the zinc atom. The ratio of the heights of these two peaks corresponds nicely to the ratio between the atomic numbers of the two species 51:30, which in turn reflects the highest amplitudes of the scattering factors for electrons. The other two peaks found by SIR2008 were much lower (305 and 300) and were, therefore, treated as ghosts. The obtained Sb and Zn positions correspond well to those of the known structure [12], with an error in atomic positions of 0.02 Å for Sb and 0.05 Å for Zn.

$CaCO_3$ (calcite)

The trigonal $CaCO_3$ (calcite) structure was selected to demonstrate a structure solution for a material with higher structural complexity, since it includes both heavy (calcium) and light (oxygen, carbon) atoms. The standard material we used consists of agglomerated crystals with a size of 200 to 400 nm, Figure 4a. It should be mentioned that such crystals are considered thick and not suitable for conventional electron diffraction studies, as dynamic diffraction effects are expected to "spoil" the intensities of the reflections significantly.

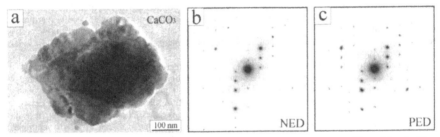

Fig. 4: Image of a typical $CaCO_3$ agglomerate (a) and diffraction pattern from ADT without (b) and with precession of the primary electron beam (c).

The ADT data was collected from the same single crystal in both the NED and PED mode with a probe size of 70 nm in diameter within a tilting range of 110°. In total 111 diffraction patterns were acquired for both tilt series. The unit cell parameters were calculated from the NED data (Figure 4b), which typically shows "better defined" diffraction spots. Due to integration over a part of reciprocal space, PED data contain errors in the position of the spots. Besides, PED patterns frequently exhibit splitting of reflections, which is partly due to the nano diffraction mode [27], Figure 4c. The lattice parameters that are determined from PED data possess, therefore, larger errors. The lattice parameters calculated from NED data were applied to the PED data set in order to index all reflections. Two intensity data sets were extracted from the NED and PED patterns, both containing 106 independent reflections.

For the NED data, the internal Rsym factor was extremely high: 72.89 %. The solution with the best residual of 37.22 % (figure of merit of 2.50) did not match the expected structure. On the other hand, the solution with the highest figure of merit (5.47, residual value 42.94 %) resembled the calcite structure, with a "wrong atomic species assignment" due to incorrect peak heights of the electrostatic potential. This solution showed additionally a ghost peak with a relatively high intensity.

The PED data set had a significantly lower internal Rsym factor of 23.4 %. The solution with the best figure of merit also had the lowest residual: 25.15 %. All atoms were found in one step at the expected positions. The Ca peak showed a height of 56, the O peak a height of 28, and the C a height of 19. The heights of these observed peaks correspond nicely to the atomic numbers of the species (56:28:19 versus 20:8:6). The first ghost peak appearing in the structure had a height of 8. As all atoms of calcite are placed on special positions, their coordinates are strictly defined. For the only partially free atom (i.e. oxygen), the displacement from the expected position was less than 0.01 Å.

BaSO₄

The orthorhombic $BaSO_4$ structure has a complexity similar to calcite as it contains both heavy and light atoms. Details of the structure solution based on ADT data have been already reported in ref. [26]. The study showed again that for an *ab initio* structure solution by direct methods, PED data are superior to NED data. Both data sets were collected in a tilt range of 120°. Again the NED data set had an internal Rsym factor significantly higher than the corresponding PED data. For NED the solution with the best residual did not correspond to that with the highest figure of merit and did not contain all oxygen atoms, whereas the heavier barium and sulfur atoms were correctly identified. Using PED data, all atoms were found in one step by SIR2008. In both cases, barium and sulfur corresponded to the expected positions with an error less than 0.05 Å, while for oxygen atoms the misplacements were bigger.

Zn₈Sb₇

The crystals of Zn_8Sb_7 were very small (i.e. 20 to 100 nm in diameter) and agglomerated with other side product phases from the same synthesis. The crystals belonging to the correct phase were unambiguously identified by ADAP 3D reciprocal volume reconstruction and cell parameters determination. The lattice parameters did not match with any known phase of $Zn_{1+x}Sb_1$ compounds, as they seemed to resemble a hexagonal arrangement with a = 15.5Å and c = 7.8Å. The analysis of the PED intensity summery revealed a non-hexagonal and non-monoclinic distribution of intensities, later confirmed with low index PED single crystal patterns.

The structure was solved from a PED data set obtained by merging two tilt series collected from the same crystal with tilt axes rotated by 90°. The number of reflection collected is impressive if compared to electron diffraction data collected by manual tilt. The internal Rsym factor is particularly low, probably due to the small volume of the analyzed crystal. A first solution in space group P1 revealed the presence of an inversion center. A second solution in space group P-1 gave the full structure. The first 14 peaks corresponded to antimony atoms while the following 16 peaks correspond to zinc. While the antimony atoms resemble a hexagonal arrangement that is reminiscent of other $Zn_{1+x}Sb_1$ structures, the arrangement of zincs breaks the symmetry down to a triclinic space group [27].

Charoite

Charoite possesses a complex framework silicate (tectosilicate) structure that is similar to a zeolite. Although this mineral is known since the last century, its structure solution has not been obtained before because it typically crystallizes in nano-fibers that are associated with other minerals. A tentative model of the structure was proposed on the basis of X-ray powder diffraction data and HRTEM [13]. The volume of the cell was assumed to be around 4500 Å³ and more than 80 atoms were assumed to be in the asymmetric unit. Contradictory results from single crystal electron diffraction pattern data made it impossible to define unambiguously the space group and the lattice parameters.

Employing ADT, it was possible to define two polymorphs with monoclinic cell and different values of the lattice parameter beta. A structurally undisturbed small single fiber (200 nm in diameter) of one of the polymorphs was selected for *ab initio* structure investigation. A very comprehensive data set was obtained merging two tilt series that were collected on the same crystal with tilt axes at 90° to each other. NED data was used to accurately define the lattice parameters. Only PED data was used for reflection intensity integration because NED data had already revealed less reliable results for much simpler structures. 8495 reflections were acquired, 2878 of which were symmetrically independent. The internal

Rsym factor was particularly good. The structural resolution was limited by the camera length used in order to correctly integrate the densely spaced spots along the a* direction (d_{100} = 32.08 Å).

The unknown structure was solved *ab initio* by SIR2008, which was able to detect almost all of the atoms in one step. 75 of the 79 atoms composing the framework of charoite were identified in one step. All of the calcium, sodium and silicon atoms were correctly localized, while 47 of the total 51 oxygen atoms were found. The content of the channels was not taken into account as the atoms inside of them have usually a less than complete occupancy. The so found model was then refined in ShelX97 [24]. The first Fourier map revealed the 4 missing oxygen atoms of the framework, all the alkali atoms, and water molecules inside the channels. A tentative least square refinement was finally performed in order to refine the occupancies of the atoms and to calculate the final composition.

2-(4-benzamido-cinnamoyl)-furane (NLO)

The non-linear optical active material 2-(4-benzamido-cinnamoyl)-furane [15] crystallizes as small facetted platelets with typical size of 100-1000 nm. The structure was solved by single crystal X-ray diffraction and consists of only light atoms. As other organic compounds, this material is very sensitive to the electron beam and can remain intact only for a few seconds under conventional TEM illumination conditions. Electron diffraction data that were collected with manual tilt (at low index zone axes) always showed a high amount of dynamic scattering.

Up to now, ADT is the only technique able to obtain enough structural information from such materials for *ab initio* structure analyses. The total tilt range of 60° was limited by the beam sensitivity of the material. The collected 1411 reflections were reduced to 428 based on the orthorhombic symmetry. The *ab initio* structure found by SIR2008 was not complete but most of the features were recognizable. Work is ongoing on Fourier refinement of this "preliminary" structural solution in order to obtain the complete structure.

Employing the ADT data set, it was possible to solve the structure using the simulated annealing procedure implemented in the program Endeavour by Crystal Impact [28]. Starting from pure diffraction information, the software was able to adjust the torsion angles and place the molecule inside the unit cell using no *a-priori* information about energy minimization. The average displacement from expected positions is 0.37 Å, of which the four highest deviations (0.6-0.5 Å) are caused by differences in torsion angles of two phenyl rings. Despite of the low reflex/parameter ratio of 4.45 the structure refinement using SHELX97 turned out to be stable even without rigid body constraints, which would not have been possible using manually collected data.

Table 2: Details of obtained structure solution, residuals refer to structure solution without refinement.

Phase	Zn_1Sb_1	$CaCO_3$ calcite		$BaSO_4$		Zn_8Sb_7		Charoite	NLO
Space group	Pbca	R-3c		Pnma		P-1		$P2_1/m$	$Pca2_1$
Volume (Å³)	438	368		347		1623		4542	1588
Atoms in asymmetric unit	2	3		5		30		89	24
Tilt range (°)	60	110		120		60+51		120+95	60
Resolution limit (Å)	1.1	0.7		0.7		0.8		1.1	1.1
Reflection coverage	70%	97%		82%		42%		97%	72%
Data type	NED	NED	PED	NED	PED	NED	PED	PED	PED
Total found reflections	388	630	701	1920	1959	4585	4613	8508	1411
Independent reflections	106	106	106	357	355	2689	2699	2878	428
Rsym	30.60%	72.89%	23.40%	48.58%	15.28%	35.72%	15.59%	13.32%	25.10%
U isotropic (Å²)	0.037	0.039	0.011	0.024	0.016	0.013	0.032	0.039	0.017
Residual	27.10%	42.94%	25.15%	42.15%	26.88%	-	34.09%	23.10%	46.47%

DISCUSSION

ADT acquisition comprises a quick and straightforward method for 3D electron diffraction data acquisition with no need for a time consuming manual crystal pre-orientation. The resolution of the data is typically below 1 Å. In some cases, the resolution was reduced because a higher camera length had to be employed in order to separate and index reflections properly for materials with large lattice constants. A larger area CCD camera (at the moment we utilize a standard 1k CCD) should solve this kind of problem.

In practice, a tilt series over 60° total tilt range (for instance ±30°) is sufficient for reliable unit cell parameter determinations. Once the geometrical corrections are employed, the software routines determine the unit cell parameters with a very high accuracy. Thus, the lattice parameters found are very reliable even for triclinic structures. The deviation of the angular lattice parameters is typically below 0.5°. The accuracy of the cell lengths parameters suffers from a systematic error that is introduced by the change of the effective camera length due to extra focusing of the nano diffraction patterns. In the mean time, a proper correction of the camera length based on the diffraction lens excitation has been implemented.

The full integration of the diffraction volume allows for conclusions about special structural features such as crystal mosaicity and partial disorder. The detection of partial disorder effects seems to be very promising. Preliminary experiments were done with Pigment Red 170 [29]. Once the unit cell is known, the direction of the faults (i.e. the directions in which the reflections are elongated) is unambiguously determined and can be indexed.

The intensity data sets extracted from the ADT diffraction patterns are significantly more powerful compared to manually collected electron diffraction data sets. ADT data sets are particularly rich in high index reflections, which for a structure solution/refinement have a similar importance as the low index reflections. Traditionally, the high index reflections were assumed to be not so important for the application of direct methods, as they typically have lower intensity. However, direct methods use normalized structure factors, so "weak" reflections may appear "strong" after the normalization. Furthermore, the high index reflections suffer less from dynamical effects. The space group is usually detected automatically as long as the extinctions are covered by the assessed reciprocal space.

Several examples of structures solved from ADT data were described in this paper comprising different classes of materials. The first structure, Zn_1Sb_1, described has a relatively small unit cell, high symmetry and possesses only strong "scatterers". For this compound, a 60° total tilt range NED data set with a resolution of 1.1 Å was sufficient for obtaining a convincing *ab initio* structure solution. $CaCO_3$ and $BaSO_4$ are more challenging as they contain both heavy and light atoms. While with NED data the structure solutions were difficult, PED data allow for a straightforward solution in one step.

For more complex structures such as Zn_8Sb_7 and charoite, the use of PED data seems to be compulsory. Such complex structures were also far out of reach for conventional electron diffraction data acquisitions based on manual tilt procedures. The solution of charoite in particular opens up new opportunities to the application of ADT to more challenging structures such as zeolites, which are otherwise not accessible with single crystal X-ray diffraction and HRTEM imaging. The Zn_8Sb_7 structure was indeed the first structure to be solved from ADT data and its solution was confirmed by the high reproducibility of the result employing subsequently independent PED data sets. The solving of the structure of Zn_8Sb_7 was also a step forward in complexity. Although it consists of only heavy atoms, the low symmetry of the structure results in a high number of atoms in the asymmetric unit.

The use of residual R factor for electron diffraction structure solution is not as straight as for X-ray crystallography, where R values below 25 % are considered as converged structures (after refinement the R values usually drop down to below 10 %). This level is rarely reached for structure solutions based on electron diffraction data. At the moment no correction is applied to our data, so the R values are relatively high. For ADT/PED data sets a solution was considered reliable if the best residual value comprised as well the highest figure of merit. For test structures the misplacement of atomic positions in respect to those determined by single crystal X-ray structure analysis was used as a measure of reliability. Basically, for judgment of the solution of unknown structures the R factor together with the overall chemical sense of the structure should be used.

CONCLUSION

The ADT acquisition and processing method, being still under development, show already in their early stages an enormous potential. The ADT technique is at present the only approach to acquire reproducibly "rich" electron diffraction data sets from single nanocrystals with dimension ranging from a few hundreds to 20 nm. The amount of reflections (i.e. is structural information in other words) that is accessible with ADT is over one order of magnitude more complete than what is achievable with manual tilt acquisitions. Moreover, the automation allows for working with non-oriented crystals, which considerably speeds up the data acquisition and allows also for working with beam sensitive materials. The time spend on diffraction data collecting may not be a severe problem for stable inorganic material, but is definitely a limiting factor when electron diffraction data from beam sensitive organic materials or proteins are to be collected. The nanocrystal imaging is done in STEM mode, which does not inflict a high electron dose on the material due to the scanning motion of the primary beam. Data sets can be collected from extremely radiation sensitive materials such as organic hydrates (drugs) and proteins [30].

The dynamic scattering associated with ADT/PED data is considerably less than for conventional electron diffraction patterns. This is partly due to the working with high indexed zones. ADT data sets have become more and more reliable and highly reproducible as different experimental settings and solutions are tested and the critical experimental parameters are better understood. NED data sets can be used for structure solution of simple structures. Once the structural complexity of the system increases, solutions from NED data become unreliable. PED data sets are definitely superior to NED data sets in terms of structural resolution.

ACKNOWLEDGMENTS

This work has been supported by the Deutsche Forschungsgemeinschaft within its Sonderforschungsbereich 625.

REFERENCES

1. M. A. O'Keefe, Microscopy and Microanalysis 10 (Suppl 2), 972 (2004).

2. X. Zou, A. Hovmöller and S. Hovmöller, Ultramicroscopy, 98, 187 (2004); J. Jansen, D. Tang, H. W. Zandbergen and H. Schenk, Acta Cryst. A54, 91 (1998); R. Kilaas, L. D. Marks and C. S. Own, Ultramicroscopy 102, 233 (2005).

3. U. Kolb, T. Gorelik, in: Weirich, Th. et al. (Eds.), Electron Crystallography, vol. 211, Kluwer Academic Publishers, Netherlands, NATO ASI Series E: Applied Sciences, 2005, p. 421.

4. B. K. Vainshtein, Structure Analysis by Electron Diffraction, Plenum, 1964.

5. R. Vincent, P. A. Midgley, Ultramicroscopy 53, 271 (1994).; C.S.Own, System design and verification of the precession electron diffraction technique, Ph. D. Dissertation, Northwestern University Evanston Illinois, 2005 /http://www.numis.northwestern.edu/Research/Current/precessions; A. Avilov, K. Kuligin, S. Nicolopoulos, M. Nickolskiy, K. Boulahya, J. Portillo, G. Lepeshov, B. Sobolev, J. P. Collette, N. Martin, A. C. Robins, P. Fischione, Ultramicroscopy 107, 431 (2007); M. Gemmi, S. Nicolopoulos, Ultramicroscopy 107, 483 (2007).

6. T. E. Weirich, R. Rameau, A. Simon, S. Hovmöller, X. D. Zou, Nature 382, 144 –146, (1996); I. G. Voigt-Martin, Z. X. Zhang, U. Kolb, C. Gilmore, Ultramicroscopy 68, 43-59 (1997); D. L. Dorset, Structural Electron Crystallography, Plenum Press, New York, 1995.

7. U. Kolb, T. Gaelic, C. Keble, M. T. Otten and D. Hubert, Ultramicroscopy 107, 507 (2007)

8. U. Kolb, T. Gaelic, M. T. Otten, Ultramicroscopy 108, 763 (2008).

9. D. Castano Diéz, A. Seybert, A. S. Frangakis, J. Struct. Biol. 154, 195 (2006).

10. E. N. Maslen, V. A. Streltsov, N. R. Streltsova and N. Ishizawa, Acta Cryst. B51, 929 (1995).

11. S. D. Jacobsen, J. R. Smyth, J. Swope, Can Miner. 36, 1053 (1998).

12. Y. Mozharivsky, A. O. Pecharsky, S. Bud'ko, and G. J. Miller: Chem. Mater. 16, 1580 (2004).

13. I. V. Rozhdestvenskaya, T. Kogure, and V. A. Drits, Abstracts of Meeting "Crystal chemistry and X-ray diffraction of Minerals", Miass 2007, p. 48-49.

14. M. U. Schmidt, S. Brühne, A. K. Wolf, A. Rech, J. Brüning, E. Alig, L. Fink, Ch. Buchsbaum, J. Glinnemann, J. van de Streek, F. Gozzo, M. Brunelli, F. Stowasser, T. Gorelik, E. Mugnaioli and U. Kolb Acta Cryst. B65, 189 (2009).

15. U. Kolb and G. N. Matveeva, Zeitschrift für Kristallographie 218(4), 259 (2003), Special issue: Electron Crystallography.

16. T. Gorelik, U. Kolb, G. Matveeva, T. Schleuß, A. F. M. Kilbinger, J. van de Streek, in preparation.

17. Basolite A100 purchased from Sigma Aldrich 688738.

18. M. N. Tahir and W. Tremel, unpublished results.

19. Hielscher USA, Inc., 19, Forest Rd., NJ 07456, Ringwood, USA.

20. MRC: basic file format of the Medical Research Council, extended with additional headers for up to 1024 images.

21. UCSF Chimera package from the Resource for Biocomputing, Visualization, and Informatics at the University of California, San Francisco (supported by NIH P41 RR-01081).

22. E. F. Pettersen, T. D. Goddard, C. C. Huang, G. S. Couch, D. M. Greenblatt, E. C. Meng, and T. E. Ferrin, J. Comput. Chem. 25, 1605(2004).

23. M. C. Burla, R. Caliandro, M. Camalli, B. Carrozzini, G. L. Cascarano, L. De Caro, C. Giacovazzo, G. Polidori, S. Diliqi, R. Spagna: J. Appl. Cryst. 40, 609 (2007).

24. G. M. Sheldrick, Acta Crystallogr. A64, 112-122 (2008).

25. S. D. Jacobsen, J. R. Smyth and J. Swope, Can. Miner. 36 (1998) 1053

26. E. Mugnaioli, T. Gorelik, U. Kolb, Ultramicroscopy, 109, 758 (2009).

27. E. Mugnaioli, T. Gorelik, M. Panthoefer, Ch. Schade, W. Tremel and U. Kolb, A combination of electron diffraction tomography and precession applied to $Zn_{1+x}Sb$ nanophases, to be published

28. H. Putz, J. C. Schoen, M. Jansen, J. Appl. Cryst., 32, 864 (1999).

29. M. U. Schmidt, D. W. M. Hofmann, Ch. Buchsbaum, and H. J. Metz, Angew. Chem. Int. Ed., 45, 1313 (2006).

30. D. Georgieva and J.-P. Abrahams, Leiden University, unpublished results.

31. Burla, M. C., Carrozzini, B., Cascarano, G. L., Giacovazzo, C. & Polidori, G. Z. Kristallogr. 217, 629 (2002).

32. Wilson A. J. C. Acta Cryst.. 3, 397 (1950).

APPENDIX

In this section some commonly used crystallographic terms are briefly explained.

The quality of a collected data set is judged by the **internal symmetry R_{sym} factor**. The R_{sym} describes the agreement between symmetrically equivalent reflections including Friedel pairs, and reflections related through the symmetry system. Therefore the equation for R_{sym} is rather simple:

$$R_{sym} = \frac{\sum_n |F_{hkl} - F_{mean}|}{\sum_n |F_{hkl}|},$$

where both summations involve all input reflections for which more than one symmetry equivalent is averaged. F_{hkl}^2 are intensities of all reflections, and F_{mean}^2 are average intensities for equivalent reflections.

Figure of merit is calculated by SIR during the phasing procedure. Only the strong reflections are used for the phasing: 70% of reflections with the strongest structure factors F_{obs}, the rest 30% should represent weak reflections. For a given set of phases the figure of merit is calculated as a ratio between the correlation coefficient R_{obs}^2 versus squared weights for the strong reflections, and the average value of R_{calc}^2 referred to weak reflections (R represents the normalized structure factor). The denominator has the major influence on the resulting value: if it is too high (weak reflections become strong) the figure of merit is low, if the denominator is low (weak reflections remain weak after phasing), the figure of merit is high. The detailed formalism of the figures of merit can be found in [31].

Final residual value R is an estimate of agreement between the experimental data and the diffraction information calculated for a given structural model. Different weighting schemes can be used, in practice the weighting coefficient is often set to 1.

$$R = \frac{\sum_{hkl} |F_{obs} - F_{calc}|}{\sum_{hkl} |F_{obs}|},$$

where the summation is done over all reflections. It was shown that for a structure with random atomic positions R value approaches 0.828 for centrosymmetric structure and 0.586 for non-centrosymmetric structure [32]. Thus the R factor of a "wrong" solution never reaches 100%.

* Corresponding author and email: kolb@uni-mainz.de

Mater. Res. Soc. Symp. Proc. Vol. 1184 © 2009 Materials Research Society   1184-GG01-06

Solving Unknown Complex Oxide Structures by Precession El3ectron Diffraction: AgCoO$_2$, PbMnO$_{2.75}$ and LiTi$_{1.5}$Ni$_{0.5}$O$_4$

H. Klein[1], M. Gemmi[1,2], A. Rageau[1]
[1]Institut Néel, Université Joseph Fourier and CNRS, 25 av. des Martyrs, BP 166, 38042 Grenoble Cedex 9
[2]Dipartimento di Scienze della Terra 'Ardito Desio', Sezione di Mineralogia, Via Botticelli 23, I-20133 Milano, Italy

ABSTRACT

Oxides synthesized in high temperature / high pressure conditions often show complex structures and contain several phases which makes a structure solution by X-ray crystallography very difficult or even impossible. Electron crystallography can then be a powerful alternative. We show here the structure solution of 3 oxides by precession electron diffraction. The phases include a simple hexagonal structure (AgCoO$_2$), a complex monoclinic structure (PbMnO$_{2.75}$) with a quasi 2-dimensional unit cell and a complex trigonal structure (LiTi$_{1.5}$Ni$_{0.5}$O$_4$). In the last case even the positions for the light element Li were determined.

INTRODUCTION

Oxides are a vast class of materials presenting various properties used in different technological applications. You can find these materials in batteries, fuel cells, high Tc superconductors, multiferroics, etc. Many of the new phases in this field are synthesized at high temperature (HT) and under high pressure (HP). Single crystals suitable for X-ray diffraction are therefore rarely available and powder X-ray diffraction suffers from severe peak overlap. Even in the rare cases where the powders contain only one phase this hinders cell parameter determination and inhibits structure solution.

Electron crystallography has proven to be a powerful tool for structure solution in cases where X-rays fail due to the above mentioned problems. Many examples have been shown in which electron crystallography has been able to solve even complex structures correctly (for recent examples see [1-3]. Recently, the availability of the precession technique for electron diffraction has permitted to collect data suitable for use in standard crystallographic software for structure determination.

However, compared to X-ray crystallography it remains a delicate and time consuming method. Therefore it finds its real application in cases where X-rays are not sufficient to solve the structures. Prominent examples are multi-phase powders constituted of nanometer sized grains.

In this contribution we give an overview over different structures we have solved by precession electron diffraction (PED). The phases include a simple hexagonal structure (AgCoO$_2$) synthesized for potential thermoelectric properties, a complex monoclinic structure (PbMnO$_{2.75}$) produced at HT/HP in search of new multiferroic materials and a trigonal structure (LiTi$_{1.5}$Ni$_{0.5}$O$_4$) fabricated in order to enhance the performance of Li batteries.

Each of these phases presents a difficulty for X-ray powder diffraction (XRPD) structure determination. The particles of the AgCoO$_2$ powder showed a particular morphology inducing preferential orientation of the particles. With this information obtained by transmission electron microscopy (TEM) the XRPD structure solution was successful at the same time as we solved the structure by PED.

The second sample is a pure powder of $PbMnO_{2.75}$. This complex structure suffers from severe peak overlap in the XRPD pattern and the cell parameters could not be determined by this technique. In a previous work we have determined the cation positions from the treatment of a high resolution electron micrograph (HREM) [4]. Here we present the structure as solved by PED.

The sample with nominal composition $LiTi_{1.5}Ni_{0.5}O_4$ contains at least 3 different phases and the structure of the main phase is sufficiently complex to cause peak overlap in the XRPD pattern. Again, the cell parameters could not be determined by XRPD and using the cell parameters obtained by TEM no satisfactory structure model could be derived even with high resolution synchrotron XRPD data. Here we present the structure model obtained by PED containing not only the cation positions but also the positions for oxygen and the very light Li.

EXPERIMENT

Data acquisition

The synthesized samples were powders but in order to obtain particles thin enough for electron diffraction each powder was crushed in an agate mortar and a suspension of the powder in ethanol was deposited on a holey carbon film for the TEM study.

The electron diffraction was performed in a Philips CM300ST operated at 300 kV and equipped with the "spinning star" precession device of Nanomegas [5]. The precession angle was varied between 0° and 4.1° which corresponds to the maximum precession angle available on our TEM. The precession angle giving the best quality data sets and retained for the structure solution differ from one structure to the other. Diffraction patterns were recorded with a GATAN Slowscan CCD camera. Exposure times were adapted for each zone axis in order to fully exploit the dynamic range of the camera and were typically in the range of a few seconds to 10 seconds.

Diffracted intensities were extracted from the diffraction patterns using the programs of the CRISP suite for electron diffraction [6] in the case of $PbMnO_{2.75}$, $AgCoO_2$ and the program QED for $LiTi_{1.5}Ni_{0.5}O_4$ [7]. Data sets obtained from the different zone axes and symmetry equivalent reflections were merged to form a unique 3 dimensional data set for each sample (except for $PbMnO_{2.75}$ where only one zone axis was used).

AgCoO2

The $AgCoO_2$ phase was synthesized by cationic replacement in a $NH_4NO_3/AgNO_3$ flux starting from $Na_{0.75}CoO_2$ in order to enhance the thermoelectric properties of the latter. The structure of $AgCoO_2$ is hexagonal of space group $P6_3/mmc$ and cell parameters $a = 2.9$ Å, $c = 12.2$ Å as determined by selected area electron diffraction (SAED).

A total of 12 zone axes were recorded in PED: [0 3 -1], [0 5 -2], [0 2 -1], [0 1 -1], [0 1 -2], [0 0 -1], [1 2 -2], [1 2 -1], [3 6 -2], [1 -1 0], [1 -2 0] and [1 0 0]. These yielded a total of 71 reflections with a resolution of $d > 0.8$ Å, 24 of which are independent with respect to the $P6_3/mmc$ symmetry. Figure 1 shows the PED patterns of the zone axes close to [0 0 1].

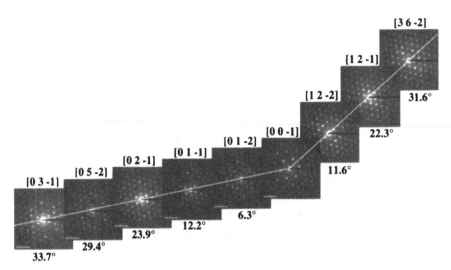

Figure 1. Precession electron diffraction patterns of zone axes close to the [0 0 -1] zone axis obtained with at precession angle of 2.4°. The indices of the zone axes are given above, the tilt angle from the [0 0 -1] zone axis below the diffraction patterns. The scale bar represents 10 nm^{-1}.

PbMnO$_{2.75}$

The PbMnO$_{2.75}$ phase was synthesized at 880 °C and 7.8 GPa. The structure is monoclinic of space group $A2/m$ and with cell parameters $a = 32.232$ Å, $b = 3.831$ Å, $c = 35.671$ Å and $\beta = 130°$ as obtained by electron diffraction [8].

A single PED pattern of the [0 1 0] zone axis was recorded for the PbMnO$_{2.75}$ with a precession angle of 2.4° (figure 2). This zone axis contains only reflections of the $h0l$ type. Therefore, only information about the x and z coordinates of the atoms can be obtained from this pattern. This is sufficient, however, since the y coordinate can be deduced by comparison to the related PbMnO$_3$ perovskite phase since the small b parameter and the space group only allow atoms to be located at $y = 0$ or $y = 0.5$.

For PbMnO$_{2.75}$ all reflections with $d > 10$ Å where the diffuse scattering around the transmitted beam gives considerable intensity were ignored for the structure determination. The resolution was limited to reflections with $d > 0.8$ Å. This yielded a set of 1965 $h0l$ reflections 991 of which were independent.

LiTi$_{1.5}$Ni$_{0.5}$MnO$_4$

The sample with nominal composition LiTi$_{1.5}$Ni$_{0.5}$O$_4$ was synthesized for use in Li-batteries by solid state reaction starting from lithium carbonate, nickel acetate and titanium oxide. The majority phase in this powder containing 3 phases was determined by electron diffraction to be of

Figure 2. Precession electron diffraction pattern of the [0 1 0] zone axis of $PbMnO_{2.75}$ obtained at a precession angle of 2.4°. The directions of the reciprocal lattice vectors **a*** and **c*** are shown by arrows. Due to the complex structure most reflections are too weak to be seen on this pattern.

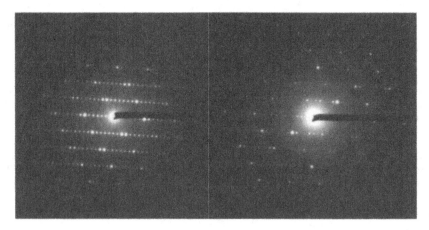

Figure 3. Comparison of the [1 0 0] zone axis of the $LiTi_{1.5}Ni_{0.5}O_4$ structure obtained in SAED (left) and PED (right). Note how different the relative intensities of the reflections in the two modes are.

trigonal symmetry with space group *P3c1* or *P-3c1* and cell parameters $a = 5.05$ Å, $c = 32.5$ Å (hexagonal setting). PED data were extracted from 10 different zone axes yielding the intensities of 105 independent reflections. Figure 3 shows the [1 0 0] zone axis recorded in SAED (left) and in PED (right). The comparison to the two patterns clearly shows the importance of acquiring data in PED if one wants to solve an unknown structure.

Data treatment

In the case of $AgCoO_2$ and $PbMnO_{2.75}$ the samples were not particularly thin. Even though the precession technique does reduce the multiple scattering in these cases, electron diffraction still obeys dynamical theory. In the framework of the 2 beam approximation of dynamical theory the measured intensities were assumed to be proportional to the structure factor amplitude [9].

The $LiTi_{1.5}Ni_{0.5}O_4$ contains no heavy elements and the sample was sufficiently thin to apply the cinematic approximation. The measured intensities were therefore assumed to be proportional to the structure factor squared.

The data were used as input for the SIR2008 [10] program for structure determination using direct methods. The only parameters used in this structure determination that didn't come directly from this work are the cell parameters, the space group and the number of atoms in the unit cell of the $PbMnO_{2.75}$ phase. These data were available from literature, but could have been determined without difficulty from electron diffraction data and crystal chemistry considerations.

In the structure determination of $PbMnO_{2.75}$ we used only one EDP corresponding to the [0 1 0] zone axis. Therefore all the measured reflections were of type $h0l$ and no information on the atom positions along the **b** axis could be obtained. Consequently, the output of SIR2008 gave for each atom a position of type $x\ 0\ z$.

Considering the small b parameter of the unit cell and the mirror plane perpendicular to **b** in the space group $A2/m$ the only possible values for y are $y = 0$ and $y = 0.5$. Since we do not have experimental evidence that allows distinguishing between the two possibilities we can only rely on a comparison with the crystal chemistry of the closely related perovskite type structure $PbMnO_3$. The first position can be chosen freely since the resulting structures will be equivalent by an origin change of the unit cell. Once the position of the first cation is chosen, its next neighbors will be attributed the same y-value if their projected distance onto the **ac**-plane is close to 3.83 Å and the y-value will be increased by 0.5 if the projected distance is considerably smaller than this value.

Since it has been demonstrated that precessed data are much more kinematical than normal SAED data [2], and since we observed a big difference between the patterns taken with precession and those without independently on the sample thickness, no attempt to solve these structures with SAED data was undertaken. Moreover we observed also that in the case of SAED it was almost impossible to merge the data from different zone axes, since the common rows of reflections displayed a different distribution of the intensity depending on the zone axis.

DISCUSSION

AgCoO₂

The structure solution obtained by SIR2008 with a reliability factor $R = 32\%$ (figure 4) contained all three independent atom positions with the highest peak in the density map corresponding to Ag and the second one to Co as should be expected. Table I compares the atomic coordinates obtained by PED with those obtained by a refinement of the structure using XRPD data. The Ag and Co atoms are in special positions where all three coordinates are fixed by the symmetry of the crystal and these positions were correctly obtained from the PED data. The oxygen is in a position where the x and y coordinates are fixed by symmetry while there is no constraint on the z coordinate. The PED solution gives the oxygen atom at a distance of 0.16 Å from the XRPD refined position.

Figure 4. Structure of AgCoO₂ obtained from SIR2008 using PED data. Silver atoms are found in between planes formed by edge sharing oxygen octahedra centered on Co.

Table I. Atom positions obtained for AgCoO₂ by PED and XRPD.

Atom	Positions obtained by XRPD			Positions obtained by PED		
	x	y	z	x	y	z
Ag	1/3	2/3	1/4	1/3	2/3	1/4
Co	0	0	0	0	0	0
O	1/3	2/3	0,08049	1/3	2/3	0,094

PbMnO₂.₇₅

The structure obtained by SIR2008 with a reliability factor $R = 41\%$ is shown in figure 5. This model only contains the cations and a clear resemblance with the perovskite structure can be seen. Regular crystallographic shear planes separate blocs of perovskite-like structures and thus determine the cell parameters a and c of the PbMnO₂.₇₅ phase. Table II gives the peak

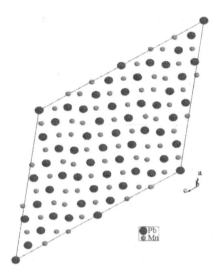

Figure 5. Projection of the cations in the structure solution for PbMnO$_{2.75}$ obtained by PED.

positions obtained by SIR2008 as well as the atom names given in [8] and the distance between the published positions and our results. As explained above we don't have any experimental information about the y coordinate of the atoms except that they have to be either $y = 0$ or $y = \frac{1}{2}$. In order to easily compare our results with the structure published by Bougerol *et al.* [8] we have chosen the same y-values as these authors.

The first 14 peaks have been correctly identified by SIR2008 as Pb and the following 15 peaks as Mn. Some of the next peaks in the output list correspond to oxygen positions in [8]. However, since not all peaks can be identified in the structural model we do not consider these positions here.

The distances between the obtained positions and those found by refinement against XRPD data by Bougerol *et al.* [8] are given in the last column of table II. The mean distance for Pb atoms is 0.21 Å, for Mn atoms it is 0.26 Å. These values show a good agreement between the two models, especially when taking into account the standard deviation of the result of the X-ray refinement of 0.1 Å for Pb and 0.3 Å for Mn.

Finally, the oxygen positions can be determined in accordance with the crystal chemistry of the perovskite phases. In the closely related phase PbMnO$_3$ the Mn atoms are in the centre of oxygen octahedra. In the **b** direction the oxygen are located at $y_{Mn} + 0.5$ and in the **ac** plane the missing oxygen atoms are placed half way between the Mn atoms.

In this way a model of the complete structure can be obtained. The atom positions are close to the real values and without a doubt close enough to do a refinement of the positions. Since the R-value of the structure solution of the electron diffraction data is rather high ($R = 0.41$), it is not reasonable to refine the positions with these data. A refinement of the positions using X-ray powder diffraction data will, however, certainly enhance the accuracy of the model.

Table II. Atomic positions for PbMnO$_{2.75}$ obtained by PED and their comparison to the XRPD refined coordinates from Bougerol et al. [8]. The last column gives the distance between each position in the two models.

atom name	Wyckhoff	x	z	name in [8]	x	z	Δ/Å
PB1	2a	0,000	0,000	Pb10	0	0	0,000
PB2	4i	0,942	0,068	Pb9	0,937	0,061	0,325
PB3	4i	0,150	0,120	Pb1	0,151	0,12	0,049
PB4	4i	0,678	0,086	Pb6	0,675	0,089	0,122
PB5	4i	0,209	0,053	Pb2	0,212	0,055	0,151
PB6	4i	0,737	0,018	Pb7	0,727	0,018	0,316
PB7	4i	0,472	0,035	Pb4	0,466	0,028	0,387
PB8	4i	0,414	0,102	Pb3	0,413	0,103	0,027
PB9	4i	0,886	0,137	Pb8	0,886	0,135	0,070
PB10	4i	0,620	0,152	Pb5	0,627	0,158	0,356
PB11	4i	0,358	0,172	Pb11	0,359	0,176	0,159
PB12	4i	0,093	0,188	Pb14	0,084	0,186	0,326
PB13	4i	0,170	0,295	Pb12	0,174	0,297	0,175
PB14	4i	0,432	0,276	Pb13	0,422	0,269	0,473
MN1	4i	0,050	0,103	Mn2	0,052	0,103	0,049
MN2	4i	0,261	0,160	Mn5	0,252	0,157	0,323
MN3	4i	0,841	0,047	Mn14	0,839	0,049	0,079
MN4	4i	0,787	0,122	Mn13	0,788	0,122	0,021
MN5	4i	0,523	0,140	Mn9	0,521	0,135	0,202
MN6	4i	0,313	0,084	Mn6	0,323	0,086	0,369
MN7	4i	0,577	0,066	Mn10	0,578	0,057	0,300
MN8	4i	0,368	0,010	Mn7	0,36	0,004	0,378
MN9	4i	0,996	0,178	Mn1	0,992	0,172	0,280
MN10	4i	0,696	0,261	Mn11	0,69	0,26	0,207
MN11	4i	0,104	0,028	Mn3	0,102	0,017	0,410
MN12	4i	0,733	0,196	Mn12	0,725	0,194	0,271
MN13	4i	0,470	0,214	Mn8	0,475	0,205	0,303
MN14	4i	0,957	0,244	Mn15	0,952	0,243	0,196
MN15	4i	0,211	0,231	Mn4	0,226	0,228	0,452

LiTi$_{1.5}$Ni$_{0.5}$MnO$_4$

Since there are two possible space groups we tried to solve the structure in the centrosymmetric space group $P\text{-}3c1$ first. The structure solution obtained ($R = 28.5\%$) is composed of layers of oxygen octahedra formed by 4 independent oxygen positions (figure 6). There are 6 independent positions for Ti and Ni occupying the centers of some of the octahedra. Discrimination between Ti and Ni proved to be difficult from theses results. Two independent positions were found for the Li ions equally occupying oxygen octahedra (table III). The solution

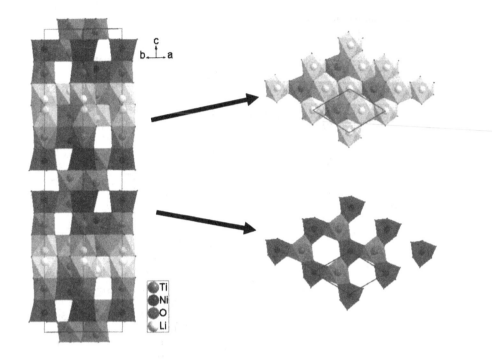

Figure 6. The structure of $LiTi_{1.5}Ni_{0.5}O_4$ obtained by PED in the centrosymmetric space group *P-3c1*. Two of the layers perpendicular to the **c** axis that are formed by edge sharing octahedra are shown on the left.

is robust against changes in the process parameters always yielding the same positions including those for Li.

However, the center of symmetry of the structure induces an occupation of the octahedra that is in violation of Pauling's second rule stating that the electrical charges of the ions are locally compensated. We have therefore decided to try a structure solution in the non centrosymmetric space group *P3c1* also. The *R* value ($R = 26.6\%$) is only slightly smaller than for the centrosymmetric space group despite the higher number of variables and the obtained structures are very similar.

It should be noted that synchrotron XRPD data collected at the high resolution powder diffraction beamline ID31 of the ESRF in Grenoble, France, did not allow us to find an *ab initio* structure solution. The best result obtained included the heavy cation and the oxygen positions but the labeling of the positions was wrong.

Table III. Atomic positions for $LiTi_{1.5}Ni_{0.5}O_4$ obtained in the centrosymmetric space group $P\text{-}3c1$ and in the non-centrosymmetric space group $P3c1$.

	centrosymmetric $P\text{-}3c1$					non-centrosymmetric $P3c1$		
Atom	**x**	**y**	**z**		**Atom**	**x**	**y**	**z**
Ti3	0	1.00000	0.14422		TI1	-1/3	1/3	0.83506
Ti4	-2/3	2/3	0.21777		TI2	0	1.00000	0.39829
TI1	-1/3	1/3	0.07746		TI3	-2/3	2/3	0.76394
TI2	-2/3	2/3	0.00860		NI1	-2/3	2/3	0.47399
Ni1	0	1.00000	0.06483		TI4	-1/3	1/3	1.03907
Ni2	-2/3	2/3	0.13164		TI5	0	1.00000	0.69502
O1	-0.26940	0.99328	0.10607		Ti6	0	1.00000	0.60787
O2	-0.33671	1.00000	1/4		Ni2	-1/3	1/3	0.62559
O3	-0.33663	0.71703	0.17815		Ni3	-2/3	2/3	0.55027
O4	-0.33355	0.60755	0.03415		Ni4	-2/3	2/3	0.67482
LI1	-1/3	1/3	0.20420		Ni5	-1/3	1/3	0.74756
Li2	0	0	0.23200		TI8	0	1.00000	0.48458
					O1	-0.26724	0.99379	0.44037
					O2	-0.70594	0.36417	1.01250
					O3	-0.68636	0.29345	0.86897
					O4	-0.25466	0.73164	0.65306
					O5	-0.35310	0.64323	0.79670
					O6	-0.04689	0.63966	0.58621
					O7	-0.69407	0.35593	0.72470
					LI1	0	1.00000	0.81784
					LI2	-1/3	1/3	0.90698
					LI3	-2/3	2/3	0.88735
					LI4	-2/3	2/3	0.83226

CONCLUSIONS

We have shown that *ab initio* structure solution of unknown oxide structures is possible for very different crystal structures. In the case of simple structures the elements can be distinguished by the intensities of the peaks in the output of SIR2008 which follow the scattering power of the elements. In the case of complex structures it might prove difficult to differentiate between cations of different nature, but they can be distinguished from the oxygen atoms.

When very heavy elements like Pb are present in the structure the precision of the measured intensities might not be sufficient to determine the oxygen positions, but the comparison with related phases and crystal chemistry often will help determine these positions.

Even though XRPD is not as efficient in structure solution as PED when it comes to powders that contain several complex phases, a combination of both techniques can prove very fruitful. A first model obtained by PED can usually be refined by XRPD data, at least when it is possible to obtain high resolution synchrotron data.

ACKNOWLEDGMENTS

It is a pleasure to thank Pierre Strobel and Hervé Mugerra for the supply of the $AgCoO_2$ and $LiTi_{1.5}Ni_{0.5}O_4$ samples and Marie-France Gorius and Catherine Bougerol for supplying the $PbMnO_{2.75}$ samples.

REFERENCES

1. *Proceedings of the Electron Crystallography School 2005, ELCRYST 2005: New Frontiers in Electron Crystallography, Ultramicroscopy* **107,** 431-558 (2007)
2. M. Gemmi and S. Nicolopoulos, *Ultramicroscopy* **107**, 483-494 (2007)
3. X. D. Zou, Z. M. Mo, S. Hovmöller, and K. H. Kuo, *Acta Cryst.* **A59**, 526-539 (2003)
4. H. Klein, *Phil. Mag. Lett.*, **85**, 569 (2005)
5. http://www.nanomegas.com
6. X.D. Zou, Y. Sukharev, and S. Hovmoeller, *Ultramicroscopy* **49**, 147 (1993).
7. D. Belletti, G. Calestani, M. Gemmi and A. Migliori, *Ultramicroscopy*, **81**, 57-65 (2000)
8. C. Bougerol, M.F. Gorius and I. Grey, *J. Solid State Chem.* **169**, 131 (2002)
9. B. K. Vainshtein, Structure Analysis by Electron Diffraction, New York: Pergamon Press. (1964).
10. http://www.ic.cnr.it/registration_form.php

Mater. Res. Soc. Symp. Proc. Vol. 1184 © 2009 Materials Research Society   1184-GG01-03

Identification of the Kinematical Forbidden Reflections From Precession Electron Diffraction

Jean-Paul Morniroli* and Gang Ji

Laboratoire de Métallurgie Physique et Génie des Matériaux, UMR CNRS 8517, USTL & ENSCL, Cité Scientifique, 59655 Villeneuve d'Ascq, France

*Corresponding author:
Jean-Paul Morniroli
Tel: 33 3 20 43 69 37
Fax: 33 3 20 43 40 40
E-mail: Jean-Paul.Morniroli@univ-lille1.fr

ABSTRACT

The visibility of the kinematical forbidden reflections due to glide planes, screw axes and Wyckoff positions is considered both on experimental and theoretical electron precession patterns as a function of the precession angle. The forbidden reflections due to glide planes and screw axes become very weak and disappear at large precession angle so that they can be distinguished from the allowed reflections and used to deduce the space groups. Contrarily, those due to Wyckoff positions remain visible and strong provided they are located on a major systematic row. This difference of behavior between the forbidden reflections is confirmed by observation of the corresponding dark-field LACBED patterns and is interpreted using the Ewald sphere and the Laue circles from the availability of double diffraction routes. This study also proves that dynamical interactions remain strong along the main systematic rows present on precession patterns.

Keywords: Electron diffraction, Electron Precession, Crystallography

INTRODUCTION

The identification of the kinematical forbidden reflections due to glide planes and screw axes is very useful for the determination of the space group of a crystal. This identification is usually easy to carry out from x-ray and neutron diffraction patterns. These patterns exhibit a kinematical behavior so that the kinematical forbidden reflections present on them are actually invisible and can be surely identified. With electron diffraction, the interactions between the incident beam and the crystal produce strong diffracted beams which may interact dynamically with the transmitted beam and with the other diffracted beams to produce double diffraction. Thus, on a Zone-Axis Pattern (ZAP), many double diffraction routes are available and some of them may produce intensity at the locations of the kinematical forbidden reflections. As a result, these forbidden reflections become visible and cannot be distinguished from the allowed reflections.

Several methods have been proposed to overcome this difficulty. The diffraction patterns can be observed on extremely thin crystal areas where a kinematical behavior prevails but this method suffers very strong limitations. Other possibilities involve the cancelation of the double diffraction routes to the forbidden reflections. On a ZAP, this situation is observed when the zone axis is perpendicular to a glide plane. In this case, all the kinematical forbidden reflections due to the glide planes are located in the Zero-Order Laue Zone (ZOLZ) and they cannot appear by double diffraction. A typical periodicity difference is then observed between the allowed reflections located in the ZOLZ with respect to the ones located in the First-Order Laue Zone (FOLZ). This feature can be used to identify the possible space groups of a crystal [1].

The cancelation of the double diffraction routes to the forbidden reflections also occur when the crystal is tilted so that only the systematic row containing the studied forbidden reflections is excited. This method is not easy to perform experimentally and confusions with weak reflections may occur.

Alternatively, when a zone axis is exactly parallel to a glide plane or exactly perpendicular to a screw axis, then the double diffraction routes to the forbidden reflections are symmetrical and they cancel two by two creating the so-called Gjønnes and Moodie lines (also called dynamical extinctions) [2] which are visible on Convergent-Beam Electron Diffraction (CBED) patterns. These dynamical extinctions can also be observed on microdiffraction ZAPs provided the required zone axis is perfectly aligned. Discontinuous lines observed on Large-Angle Convergent-Beam Electron Diffraction patterns (LACBED) are also connected with forbidden reflections [3].

A new possibility of identification of the kinematical forbidden reflection is now available with the development of electron precession proposed by Vincent and Midgley [4]. It was reported that the kinematical forbidden reflections present on Precession Electron Diffraction (PED) patterns become very weak and eventually disappear with large precession angles (about 3°) [5]. Nevertheless, some exceptions to this behavior were observed by Vincent and Midgley [4] and Own et al. [6]. For example, the {222} forbidden reflections remain visible on precession patterns from silicon specimens.

The aim of this paper is to explain the origin of these exceptions in order to deduce some rules useful for the identification of the kinematical forbidden reflections from PED patterns. To this aim, silicon, white tin and AsNb crystals were considered experimentally or theoretically. All of them are well adapted to this study since their diffraction patterns exhibit kinematical forbidden reflections due to glide planes and screw axes but also to special Wyckoff positions.

Electron precession patterns are usually interpreted by means of the Ewald sphere and Laue circles or by using dedicated kinematical [7] or dynamical simulations [8]. In the present paper, we also consider dark-field LACBED patterns taking into account the strong analogy which exists between both techniques [9].

COMPARATIVE DESCRIPTION OF THE PRECESSION AND DARK-FIELD LACBED TECHNIQUES

A comparative description between electron precession and dark-field LACBED was given in a previous reference [9]. It indicates that the integrated intensity displayed by a hkl spot on a PED pattern obtained with a precession angle α (Fig. 1a) is connected with the intensity observed along the circle drawn on a hkl dark-field LACBED pattern (Fig. 1b) which corresponds to the beam convergence α. It is clear that during a 360° precession rotation, the Bragg line present on the LACBED pattern is scanned two times. The effect of the precession angle on the hkl intensity on the PED pattern can be obtained by considering the corresponding circles on the hkl dark-field LACBED pattern.

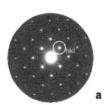
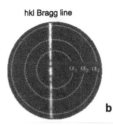

hkl Bragg line

Figure 1: Relationship between a hkl reflection on a PED pattern and its corresponding dark-field LACBED pattern.
a - PED pattern.
b - hkl dark-field LACBED pattern.

INTERPRETATION OF THE PRECESSION PATTERNS FROM EWALD SPHERE AND LAUE CIRCLES

The PED patterns can be easily interpreted by means of the Ewald sphere and the Laue circles.

Let us consider a [uvw] ZAP. In this case, all the nodes of the reciprocal lattice can be considered as stacked along parallel and equidistant (uvw)* reciprocal layers perpendicular to the zone axis as shown on Fig. 2a. Each layer is characterized by an integer index n; the layer with n=0 containing the origin O* of the reciprocal lattice. Actually, due to the small thickness of the specimens observed in electron diffraction, the nodes of the reciprocal lattice are not punctual but they are elongated and form "relrods" whose size is inversely proportional to the crystal thickness t. The Laue circles (continuous lines on Fig. 2a) correspond to the intersection of these (uvw)* layers with the Ewald sphere and they represent the geometrical loci for diffraction.

For a conventional [uvw] ZAP, the Laue circles are centered on the zone axis. Note that the intersection of the layer with n=0 is a point. If the relrods are taken into account, then each Laue circle is transformed into a Laue area (shaded area on Fig. 2a) whose width is connected with the crystal thickness. Each node of the reciprocal lattice located within these Laue areas is excited and produces a diffracted beam. A Whole Pattern made of the ZOLZ and some High-Order Laue Zones (HOLZ) is formed in that way (black nodes on Fig. 2a). For the sake of clarity, only the layer with n=0 will be considered hereafter.

For a precession pattern, the incident beam is tilted from the zone axis by the precession angle α and rotates around it. The corresponding Laue circles and areas are then shifted and they also rotate around the zone axis as shown on Fig. 2b. The precession pattern is formed in a serial way after a 360° rotation of the Laue areas. Figs. 2c to f indicate that the radius of the Laue circles increases with the microscope voltage V and with the precession angle α while the width of the Laue areas decreases with the specimen thickness t. Thus, at large precession angles and with thick specimens, the number of reflections simultaneously excited during the precession movement becomes small so that a "few-beam" behavior is encountered (Fig. 2e). For some special positions of the Laue areas a "systematic row" behavior is obtained (Fig. 2f).

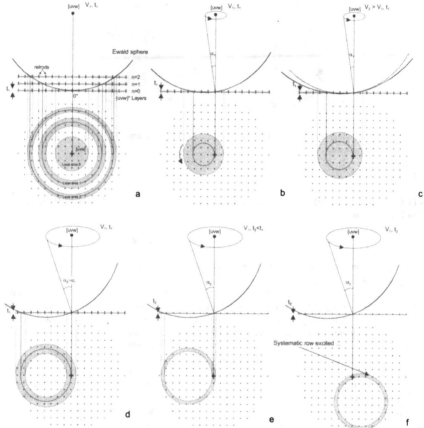

Figure 2: Interpretation of the precession patterns by means of the Ewald sphere and the Laue circles.

a - Formation of a ZAP.

b to f - Formation of a PED pattern. Effects of the voltage (b, c), the precession angle (b, d) and the specimen thickness (d, e). A "few-beam" (e) or a "systematic row" (f) behavior is encountered at large precession angles.

KINEMATICAL FORBIDDEN REFLECTIONS

Kinematical forbidden reflections correspond to hkl reflections whose structure factor equals to zero. This situation occurs when the crystal structure contains glide planes (a, b, c axial, n diagonal and d diamond glide planes), screw axes (2_1, 3_1, 3_2, 4_1, 4_2, 4_3, 6_1, 6_2, 6_3, 6_4 and 6_5 screw axes) or when the atoms occupy special Wyckoff positions in the unit cell.

The glide planes only affect reflections of the type hk0, h0l, 0kl or hhl (zonal reflection conditions). General hkl reflections are never concerned. This means that the corresponding forbidden nodes of the reciprocal lattice are all located in the (uvw)* layer of the reciprocal lattice which is simultaneously parallel to the glide plane ([uvw] is then perpendicular to the glide plane) and which contains the origin O* of the reciprocal lattice. The other (uvw)* layers with n≠0 are not affected.

The screw axes only affect h00, 0k0 or 00l nodes (serial reflection conditions) located along reciprocal rows simultaneously parallel to the screw axes and containing the origin O* of the reciprocal lattice.

On a diffraction pattern, the kinematical forbidden reflections due to the glide planes and the screw axes are always located along one, two or four systematic rows on a diffraction pattern. Depending on the nature of the symmetry elements, four different types of systematic rows containing forbidden reflections are encountered as shown on Fig. 3a. Along all of them, the forbidden reflections are regularly disposed between the allowed reflections.

The forbidden reflections due to the Wyckoff positions may affect any type of reflections including hkl reflections. Note that they only occur if all the atoms present in the unit cell are located on the same Wyckoff position. This strict condition means that these forbidden reflections are mainly observed with some pure elements (C, Si, Ge, $_\alpha$Sn with diamond structure, Co, Mg, Cd, Pb, Ti, Zr, Zn... with hcp structure, $_\beta$Sn, Pu....) or with some very simple crystal structures (ZnS, AsNb...). The corresponding forbidden reflections are also located along systematic rows on the diffraction patterns. Nevertheless, they are not arranged in a regular way as shown on the example of the {hhh} systematic row observed with silicon patterns (Fig. 3b).

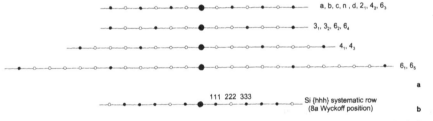

a, b, c, n , d, 2_1, 4_2, 6_3

3_1, 3_2, 6_2, 6_4

4_1, 4_3

6_1, 6_5

a

111 222 333
Si {hhh} systematic row
(8a Wyckoff position)

b

○ Kinematical forbidden reflections
● Allowed reflections

Figure 3: Arrangement of the kinematical forbidden reflections along a systematic row.
a - Kinematical forbidden nodes due to glide planes and screw axes.
b - Example of kinematical forbidden reflections due to a Wyckoff position.

DOUBLE DIFFRACTION

Double diffraction is frequently observed in electron diffraction due to the dynamical behavior of the crystal-electrons interactions. As shown on Fig. 4a, a beam diffracted at first by the $(h_1k_1l_1)$ lattice planes can be diffracted a second time by the $(h_2k_2l_2)$ lattice planes according to the double diffraction route $\vec{s_3} = \vec{s_1} + \vec{s_2}$. The three reflections $h_1k_1l_1$, $h_2k_2l_2$ and $h_3k_3l_3$ involved in the double diffraction process are connected by the relationships: $h_3 = h_1 + h_2$; $k_3 = k_1 + k_2$; $l_3 = l_1+l_2$. On a conventional ZAP, where many diffracted beams are simultaneously excited, several double diffraction routes are available and some of them, shown on Fig. 4b, can produce intensity at the location of the kinematical forbidden reflections. The situation is different on precession patterns. During the precession movement, the Laue circles rotate around the zone axis. A double diffraction route to a forbidden reflection is only available when the two allowed reflections implied in the double diffraction route and the forbidden reflection are simultaneously located within a Laue area. This situation only occurs for some special positions of the Laue area during its rotation around the zone axis like the one shown on Fig. 4c. Consequently, the possibilities of double diffraction are then strongly reduced with electron precession.

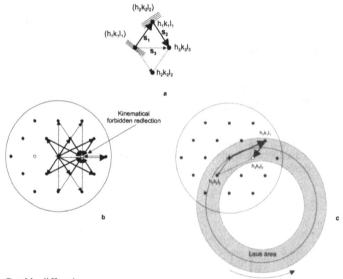

Figure 4: Double diffraction.
a - Schematic description of a double diffraction route.
b - Conventional ZAP. Several double diffraction routes to a kinematical forbidden reflection are available.
c - PED pattern. A double diffraction route to a kinematical forbidden reflection is only available when the three involved reflections $h_1k_1l_1$, $h_2k_2l_2$ and $h_3k_3l_3$ are simultaneously located within the Laue area.

EXPERIMENTAL AND THEORICAL CONDITIONS

All the experiments were performed on a Philips CM30 transmission electron microscope working at 300 kV. The patterns were recorded on a CCD camera.

The PED patterns were obtained in the microprobe mode with the Spinning Star equipment from Nanomegas. The maximum precession angle was around 3.2°. In order to probe a small specimen area, the microdiffraction mode was selected by using a nearly parallel incident beam produced by a 10 μm C_2 aperture. The incident electron beam was focused on the specimen with a spot size in the range 10-50 nm. The LACBED patterns were obtained in the nanoprobe mode with a beam convergence of about 3° and a spot size in the range 5 to 50 nm.

The silicon specimens were prepared using the tripod method.

The theoretical precession patterns from βSn and AsNb were simulated with the Jems software [8] using the following parameters: 300 kV, 3° precession semi-angle, 0.5 mrad incident beam convergence, 200 nm specimen thickness and 0.01° precession angle increment.

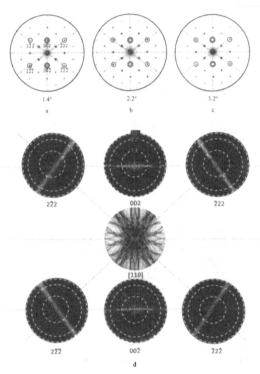

Figure 5: Silicon [110] ZAPs

a to c - PED patterns obtained at three different precession angles. The double circled spots are kinematical forbidden reflections due to the d glide planes. The circled spots are kinematical forbidden reflections due to 8a Wyckoff positions.

d - Bright-field and Dark-field LACBED patterns of the kinematical forbidden reflections.

Two different silicon ZAPs were experimentally studied: <110> and <112>. Silicon belongs to the space group $Fd\bar{3}m$ and the Si atoms are located on the 8a Wyckoff positions (000 and ¾ ¼ ¾).

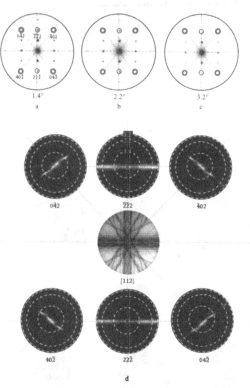

Figure 6: Silicon [112] ZAPs
a to c - PED patterns obtained at three different precession angles. The double circled and circled spots are kinematical forbidden reflections due to glide planes and 8a Wyckoff positions, respectively.
d - Bright-field and Dark-field LACBED patterns of the kinematical forbidden reflections.

<110> ZAPs

The kinematical forbidden reflections which can be observed on these ZAPs can be sorted into two types:
- the 002, 00$\bar{2}$, 006, 00$\bar{6}$... forbidden reflections (double circled reflections on Figs. 5a, b and c). These forbidden reflections are due to the presence of d diamond glide planes parallel to the {001} lattice planes.

- the $2\bar{2}2$, $\bar{2}22$, $2\bar{2}\bar{2}$, $\bar{2}2\bar{2}$, $4\bar{4}2$, $4\bar{4}\bar{2}$, $\bar{4}4\bar{2}$, $\bar{4}42$, $2\bar{2}6$, $\bar{2}26$, $\bar{2}2\bar{6}$, $2\bar{2}\bar{6}$... forbidden reflections (circled reflections on Fig. 5a, b and c). They are due to the 8a Wyckoff positions which produce extinctions on hkl reflections when h is odd or when h+k+l=4n.

The [110] experimental PED patterns (Figs. 5a, b and c) performed with three different precession angles, namely: 1.4°, 2.2° and 3.2° clearly indicate a different behaviour between the 002 and $00\bar{2}$ forbidden reflections and the $2\bar{2}2$, $\bar{2}22$, $2\bar{2}\bar{2}$ and $\bar{2}2\bar{2}$ forbidden reflections. The first ones disappear at large precession angles while the second ones remain visible and strong whatever the precession angle.

The dark-field LACBED patterns of the corresponding forbidden reflections are also given on Figs. 5d. The dotted circles drawn on these patterns correspond to the three precession angles. They confirm that the intensities of the 002 and $00\bar{2}$ reflections are strong at 1.4° and rapidly decrease at larger precession angles. On the contrary, the $2\bar{2}2$, $\bar{2}22$, $2\bar{2}\bar{2}$ and $\bar{2}2\bar{2}$ Bragg lines exhibit visible intensities for all values of the precession angles.

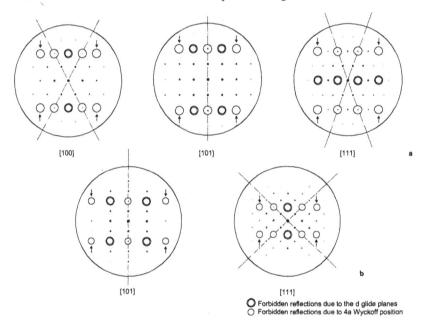

[100]   [101]   [111] a

[101]   [111] b

○ Forbidden reflections due to the d glide planes
○ Forbidden reflections due to 4a Wyckoff position

Figure 7: Theoretical PED patterns with 3° precession angle from $_\beta$Sn (a) and AsNb (b) crystals. Some intensity is observed for the forbidden reflections due to the Wyckoff positions which are located on the systematic rows indicated by a continuous line.

<112> ZAPs

The kinematical forbidden reflections present on these ZAPs can also be sorted into two types:

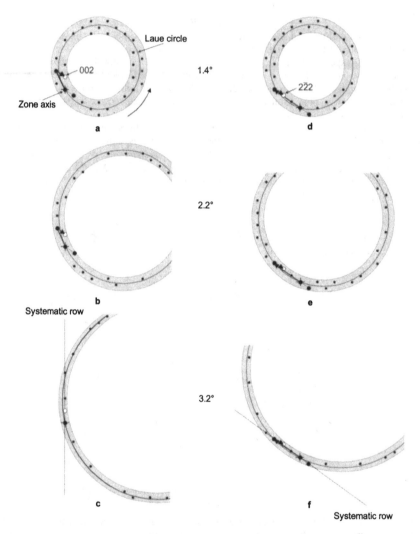

Figure 8: Availability of double diffraction routes to the 002 (a, b and c) and $2\bar{2}2$ (d, e and f) kinematical forbidden reflections of silicon as a function of the precession angle.

- the $40\bar{2}$, $0\bar{4}2$, $\bar{4}02$ and $04\bar{2}$ forbidden reflections due to the d glide planes parallel to the {001} lattice planes (double circled reflections on Figs. 6a, b and c).
- the $22\bar{2}$, $\bar{2}\bar{2}2$, $2\bar{6}2$, $6\bar{2}\bar{2}$, $\bar{2}6\bar{2}$, $\bar{6}22$... forbidden reflections due to the 8a Wyckoff position (circled reflections on Figs. 6a, b and c).

On the experimental PED patterns (Figs. 6a, b and c), the $40\bar{2}$, $0\bar{4}2$, $\bar{4}02$ and $04\bar{2}$ reflections disappear at large precession angles while the $22\bar{2}$ and $\bar{2}\bar{2}2$ reflections remain visible for all precession angles. The same behaviour is also clearly visible on the corresponding dark-field LACBED patterns (Figs. 6d).

THEORETICAL RESULTS

In order to confirm the experimental results obtained from the silicon specimen, dynamical precession simulations were also carried out with the Jems software [8] from $_\beta$Sn and AsNb specimens. These two crystals were selected because their atoms are located at special Wyckoff positions which produce additional kinematical forbidden reflections to the ones connected with glide planes and screw axes.
- $_\beta$Sn belongs to the space group $I\,4_1/a\,md$ and the Sn atoms occupy the 4a Wyckoff positions (000 and 0 ½ ¼) producing the following additional reflection conditions on hkl: l=2n+1 or 2h+l=4n.
- AsNb belongs to the space group $I4_1md$ and both the As and Nb atoms are located on 4a Wyckoff positions (00z and 0 ½ z+¼) with z=0.416 for As and 0 for Nb. The additional reflections conditions on hkl are also l=2n+1 or 2h+l=4n.
The simulated $_\beta$Sn and AsNb zone-axis PED patterns shown on Fig. 7 were selected because they simultaneously display kinematical forbidden reflections due to glide planes and screw axes (double circled reflections) but also to the 4a Wyckoff positions (circled reflections). With a 3° precession angle, the forbidden reflections due to the glide planes or to the screw axes are actually invisible on these patterns while those due to the 4a Wyckoff positions are visible and strong provided they are located on a major systematic row (continuous lines on Fig. 7).

INTERPRETATIONS

In order to interpret the difference of behaviour between the forbidden reflections due to the glide planes, the screw axes or the special Wyckoff positions, let us consider, for example, the visibility of the 002 and $2\bar{2}2$ reflections from silicon as a function of the precession angle (Fig. 5). The first studied reflection is due to the d glide plane and the second one to the 8a Wyckoff position.
For the 002 reflection, double diffraction routes to this reflection are only available at low precession angles for the special Laue circle position shown on Figs. 8a and b. At 3.2° (Fig. 8c), where a "systematic-row" behaviour is observed, double diffraction routes to the forbidden reflection are no longer available. This feature explains why this forbidden reflection becomes invisible at large precession angle. The situation is different for the $2\bar{2}2$ reflection (Figs. 8d, e and f). Even at large precession angles, double diffraction routes to the forbidden reflections are available along the systematic row which contains the forbidden reflection.
The same explanation can be given for all the other forbidden reflections present on the experimental and theoretical patterns on Figs. 5 to 7. Thus, at large precession angle, the double diffraction routes mainly occur between allowed reflections located along a systematic row.
What about the forbidden reflections due to the Wyckoff positions which also disappear at large precession angle (see the arrowed reflections on Figs 7a and b). This behaviour is connected with the fact that these reflections are located on a systematic row with low symmetry

47

(dotted lines on Fig. 8) along which the allowed reflections are too spaced to produce double diffraction routes.

A simple rule may be deduced from these observations. The kinematical forbidden reflections disappear at large precession angles provided they are disposed regularly between allowed reflections along a systematic row. This is the case of the forbidden reflections due to glide planes and screw axes (Fig. 3a). The forbidden reflections due to Wyckoff positions never satisfy this rule (Fig. 3b).

This study proves that electron precession is well adapted to the identification of the forbidden reflections due to glide planes and screw axes. As a matter of fact, these reflections are useful when dealing with space group identification. The best experimental conditions are encountered with a large precession angle (about 3°), a high voltage and a thick specimen.

CONCLUSIONS

Kinematic forbidden reflections due to glide planes, screw axes and Wyckoff positions were studied experimentally and theoretically on silicon, $_\beta$Sn and AsNb PED and/or dark-field LACBED patterns performed at different precession or convergence angles (up to 3°). Observations of these patterns indicate that the forbidden reflections due to glide planes and screw axes disappear at large precession or convergence angles while those due to Wyckoff positions remain visible for all angles provided they are located along a major systematic row.

This behavior is interpreted by means of the Ewald sphere and the Laue circles which reveals that a "few-beam" or a "systematic-row" behavior prevail at large precession angles. They also indicate that the forbidden reflections disappear at large precession angle because the double diffraction routes to them are canceled along the systematic row where the forbidden reflections are located. This situation only happens when the forbidden reflections are regularly disposed between the allowed ones. This is the case for the forbidden reflections due to glide planes and screw axes. The forbidden reflections due to Wyckoff positions are not disposed in that way and then, they remain visible for all precession angles unless they are located on a low symmetry systematic row. As a result, the kinematical forbidden reflections due to glide planes and screw axes can be identified from the allowed reflections on PED patterns. This identification is especially useful for space group determination.

This study also proves that dynamical effects remain strong along the major systematic rows present on PED patterns. This feature should be taken into account when dealing with structure determination based on intensity measurements.

REFERENCES
1. J.P. Morniroli and J.W. Steeds, Ultramicroscopy 45, 219 (1992).
2. J. Gjønnes and A.F. Moodie, Acta Cryst. 19, 65 (1965).
3. J.W. Steeds, Microscopia Elettronica in Trasmissione e Techniche di Analisi di Superficici nella Scienza dei Materiali, Editione Enea, Roma, 1986
4. R. Vincent and P.A. Midgley, Ultramicroscopy 53, 271 (1994).
5. J.P. Morniroli, A. Redjaïmia and S. Nicolopoulos, Ultramicroscopy 107, 514 (2007).
6. C.S. Own, L.D. Marks and W. Sinkler, Acta Cryst. A62, 434 (2006).
7. J.P. Morniroli, Electron Diffraction, http://www.univ-lille1.fr/lmpgm/
8. P.A. Stadelmann, Jems, http://cimewww.epfl.ch/people/stadelmann/jemsWebSite/jems.html
9. J.P. Morniroli, Microscopy and Microanalysis, 13, 126-127 (2007)

Mater. Res. Soc. Symp. Proc. Vol. 1184 © 2009 Materials Research Society 1184-GG03-07

Structural Fingerprinting of Nanocrystals: Advantages of Precession Electron Diffraction, Automated Crystallite Orientation and Phase Maps

Peter Moeck[1], Sergei Rouvimov[1], Edgar F. Rauch[2], and Stavros Nicolopoulos[3]

[1] Nano-Crystallography Group, Department of Physics, Portland State University, Portland, OR 97207-0751 & Oregon Nanoscience and Microtechnologies Institute
[2] SIMAP/GPM2 Laboratoire, CNRS-Grenoble INP, BP 46 101 rue de la Physique, 38402 Saint Martin d'Hères, France
[3] NanoMEGAS SPRL, Boulevard Edmond Machterns No 79, Saint Jean Molenbeek, Brussels, B-1080, Belgium

ABSTRACT

Strategies for the structurally identification of nanocrystals from Precession Electron Diffraction (PED) patterns in a Transmission Electron Microscope (TEM) are outlined. A single-crystal PED pattern may be utilized for the structural identification of an individual nanocrystal. Ensembles of nanocrystals may be fingerprinted structurally from "powder PED patterns". Highly reliable "crystal orientation & structure" maps may be obtained from automatically recorded and processed scanning-PED patterns at spatial resolutions that are superior to those of the competing electron backscattering diffraction technique of scanning electron microscopy. The analysis procedure of that automated technique has recently been extended to Fourier transforms of high resolution TEM images, resulting in similarly effective mappings. Open-access crystallographic databases are mentioned as they may be utilized in support of our structural fingerprinting strategies.

INTRODUCTION

This paper outlines structural fingerprinting strategies of nanocrystals in a TEM on the basis of PED patterns [1-7]. To appreciate the advantages of this diffraction mode for structural fingerprinting, its basics will be discussed in the main part of this paper. There is also structural fingerprinting on the basis of high resolution (HR or lattice-fringe) TEM images [8-10]. That method and its unique advantage for structural fingerprinting, i.e., the possibility to extract structure factor phase angles will, however, not be discussed here since a comprehensive description is published in open access [8]. We will accordingly deal only in passing with the possibility that one can obtain highly reliable crystal orientation & structure maps from HRTEM images [11] by an extension of the automated processing of single-crystal PED patterns [4,7].

Searching for structural information that is extracted by our structural fingerprinting strategies in crystallographic databases and matching it with high figures of merit to that of candidate structures allows for highly discriminatory identifications of nanocrystals, even without additional chemical information as obtainable in analytical TEMs. As an alternative to commercial databases, one may use open-access databases, which provide together some 100,000 crystal structure data sets (that include atomic coordinates, unit cell parameters, and the space group). The major open-access crystallography databases that can be utilized for structural fingerprinting from PED patterns will be mentioned at the end of this paper.

Since our structural fingerprinting strategies utilize structure factor modulus information that is extracted from the intensity of the reflections, they are restricted to nanocrystals. The nanocrystals need to be sufficiently thin so that kinematic or quasi-kinematic approximations to the scattering of "fast" electrons are applicable. Fast means here some 50 to 80 % of the speed of light, corresponding to acceleration voltages of 100 to 300 kV. Kinematic and quasi-kinematic approximations to the scattering of fast electrons form the basis of structural electron crystallography [12-15] and will be discussed next.

KINEMATIC AND QUASI-KINEMATIC APPROXIMATIONS TO THE SCATTERING OF FAST ELECTRONS

The two-beam dynamical electron scattering theory suffices frequently for structural fingerprinting and structural electron crystallography. As will be shown below, for vanishing crystal thickness, the predictions of the two-beam dynamical scattering theory closely approach the predictions of the kinematic theory. We consider this theory in its low thickness limit, therefore, as a quasi-kinematic theory.

The conceptual basis of the kinematic theory is single scattering of electrons by the electrostatic potential of a crystal. The electrons are scattered out of the primary beam into diffracted beams while the former is negligibly attenuated. This is an idealized case for the scattering of electrons (while it typically suffices for the scattering of X-rays by crystals that are composed of mosaic blocks in the millimeter range). For nanometer sized crystals one can, however, reliably base crystallographic analyses by means of electron scattering on the low thickness limit of the two-beam dynamical theory and correct for the primary extinction effects to the reflection intensities. The physical process of electron diffraction can be described mathematically by a Fourier transform. Following ref. [15] and using some of its notation, the Fourier coefficients, Φ_{hkl}, of the electrostatic potential $\phi(x,y,z)$ are given by the relation

$$\Phi_{hkl} = \int_{\Omega} \phi(x,y,x) \cdot \exp 2\pi i (hx + ky + lz) \cdot dxdydz \tag{1},$$

where the dimension of Φ_{hkl} is volts times cube of length; $\Omega = \vec{a} \cdot (\vec{b} \times \vec{c})$ is the volume of the unit cell; $\vec{a}, \vec{b}, \vec{c}$ are the basis vectors; h,k,l are the Miller indices; and x,y,z are the fractional coordinates of atoms in the unit cell. These Fourier coefficients represent electron waves that are scattered by the electrostatic potential of the crystal in directions that are defined by Bragg's law and recorded a large distance away from the crystal. When the unit cell volume is given in nm^3, the relation $|F_{hkl}| \approx 20.8877 \cdot |\Phi_{hkl}|$ is valid for structure factor moduli in Å. The structure factors, F_{hkl}, (and the Fourier coefficients of the electrostatic potential) are complex entities, i.e. possess a modulus and a phase angle. The former can be calculated from the relation

$$F_{hkl} = \sum_j f_j \cdot f_T{}^j \cdot \exp 2\pi i (hx_j + ky_j + lz_j) \tag{2},$$

where f_j is the atomic scattering factor (for electrons), and $f_T{}^j$ is the respective temperature factor for all j atoms in the unit cell. (Note that temperature factors are much less important for electrons than for X-rays. This is because the atomic scattering factors fall for electrons off with $(\sin\Theta/\lambda)^2$.)

For an **ideal** single crystal, the two-beam dynamical diffraction theory gives the intensity of a diffracted electron beam, I_{hkl}, (i.e. of a reflection in an electron diffraction spot pattern) by the relation

$$I_{hkl} = I_0 S \cdot Q^2 \frac{\sin^2\{t[(\pi T)^2 + Q^2]^{0.5}\}}{(\pi T)^2 + Q^2} \tag{3},$$

with I_0 as intensity of the primary electron beam; S as area of the crystal that is illuminated by the primary electron beam; λ as relativistic electron wavelength; $Q = \lambda/\Omega \cdot |F_{hkl}|$ as an entity that is

50

proportional to the respective structure factor modulus; t as crystal thickness; and T as extension of the reciprocal lattice node.

The respective prediction of the kinematic theory for an **ideal** single crystal is:

$$I_{hkl} = I_0 S \cdot Q^2 \frac{\sin^2\{t \cdot \pi T\}}{(\pi T)^2} \qquad (4).$$

If Q is much smaller than pT, relation (3) can be well approximated by relation (4). Because T is inversely proportional to the size of the crystal, it becomes larger the smaller the nanocrystal gets. In other words, for sufficiently thin crystals, the two-beam dynamical diffraction theory is well approximated by the kinematic theory. Note that the nature of the electron scattering phenomena is revealed in relations (3) and (4), but one does not base structural fingerprinting strategies that employ the reflection intensities directly on them. For that, these theoretical relations need to be modified by Lorentz factors that refer to the **real** prevailing experimental conditions. The ratio of the integrated scattered beam intensity to the initial beam intensity received by a **real crystalline sample** from the primary beam in a **real** electron scattering experiment is called "integrated coefficient of reflection" [15]. It is in the kinematic theory given by the relation

$$\frac{I_{hkl}}{I_0 S} = Q^2 \cdot t \cdot L \qquad (5),$$

where L is a Lorentz factor and possesses the unit of length. Analogously to their counterparts in X-ray diffraction, Lorentz factors account for the physical particulars (including the relative time intervals) of the intersections of the Ewald sphere with the shape transform of the nanocrystals at the accessible reciprocal lattice points.

The nature of the Lorentz factor differs from experimental set up to experiment set up, i.e. with both the diffraction technique and the crystalline sample type. Within a certain diffraction technique and crystalline sample type, the Lorentz factor varies only quantitatively [14,15]. Making L smaller than t and/or the reciprocal value of Q by choice of certain experimental parameters of a diffraction technique or by choice of the selection of a certain crystalline sample reduces the integrated coefficient of reflection so that structural fingerprinting may proceed within the frameworks of kinematic or quasi-kinematic electron scattering theories.

The electron wavelength, thickness and structure of the nanocrystal, as well as the volume of its unit cell are fixed in a typical experiment, but are also parameters that determine how well the two-beam dynamical diffraction theory will be approximated by the kinematic theory. It is, therefore, helpful to introduce a "range of crystal thicknesses, structure factor moduli, electron wavelengths, and unit cell volumes" in which a nanocrystal diffracts quasi-kinematically. Following Boris K. Vainshtein, refs. [14,15], the relations

$$Q \cdot t' = \frac{\lambda}{\Omega} \cdot |F_{hkl}| \cdot t' \le 1 \approx 1 \qquad (6),$$

where t' has the meaning of a "critical thickness range" can be used as an evaluation criterion for the gradual transition from the kinematic theory to the dynamical two-beam theory.

As first proposed by Blackman [16], corrections for primary extinction effect can be made in the two-beam approximation as long as the experimental diffraction technique provides an integration of the reflection intensities. This is obviously the case for powder patterns [14,15] and also to some extent for PED patterns [17,18]. In addition, one may deal with systematic n-beam interactions of selected systematic rows, e.g. for *(h00)*, *(hh0)*, and *(hhh)* reflections that are higher orders *(n = 2, 3, ...)* of strong reflections with $h = 1$ or 2, by means of the "Bethe dynamic potentials" [19]. As in the Blackman correction, no exact prior knowledge of the crystal thickness or orientation is needed to apply this correction.

PRECESSION ELECTRON DIFFRACTION

The single-crystal PED technique is formally analogous to the well known (single-crystal) X-ray (Buerger) precession technique. It utilizes, however, a precession movement of the primary electron beam around the microscope's optical axis rather than that of a zone axis of a single crystal around a fixed primary X-ray beam direction. Due to the much larger radius of the Ewald sphere, the precession angles are in PED only a few degrees, i.e. an order of magnitude smaller than in X-ray precession. The primary electron beam can be either parallel or slightly convergent and its precession creates a hollow illumination cone which has its vertex on the crystalline sample. The primary electron beam and the diffracted beams are de-scanned (after they have left the exit face of the nanocrystal) in such a manner that stationary diffraction patterns are obtained on the (stationary) viewing screen of the TEM [17,18].

Figures 1a to 1c illustrate effects of a precessing primary electron beam on the [110] zone-axis diffraction pattern of a silicon nanocrystal. The projection of the precession movement of the primary electron beam (around the optical axis of the microscope onto the viewing screen of the TEM) in direct space is equivalent to the rotation of the so called "Laue circle" in reciprocal space, Fig. 1a. The circumference of the Laue circle represents the locations of intersections of the Ewald sphere with zero-dimensional nodes of the reciprocal crystal lattice. Due to the small size of nanocrystals, these nodes are extended in three dimensions by the Fourier transform of the crystal shape function. Individual reflections of real nanocrystals that are located close to the circumference of the Laue circle are, therefore, excited more or less sequentially, Fig. 1a, while the Laue circle as a whole rotates around the central *000* spot of the diffraction pattern. Larger precession angles are more effective in exciting reflections sequentially.

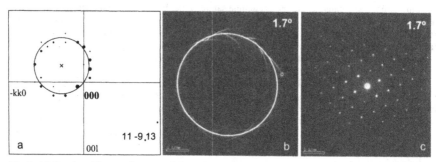

Fig. 1: (a) Sketch to illustrate the sequential creation of a single-crystal PED pattern by rotation of the Laue circle around the origin of the reciprocal lattice; Silicon, 10 nm thickness, [110], 200 kV, 1.7° precession angle. Note that in addition to reflections of the zero order Laue zone, precession with a sufficiently large angle excites also reflection from the second order Laue zones, e.g. (11 -9 13). All reflections of the first order Laue zone are for this zone axis orientation of Si kinematically forbidden. **(b)** Experimental PED pattern in the "just-precessed" mode; Silicon, 40 nm approximate thickness, [110], 200 kV, 1.7° precession angle. The circle of high intensity represents the precessing primary electron beam. The arcs of lower intensity are the traces of the intersection of the individual Bragg reflections with the precessing Ewald sphere. **(c)** Experimental PED pattern in the "properly de-scanned" mode; all parameters as in (b). Note that a stationary spot diffraction patterns results from the integration of all arcs of low intensity in (b).

The locations of the individual reflections on the diffraction pattern appear stationary as a result of the proper de-scanning of the diffracted electron beams, Fig. 1c, but their intensities vary while the Laue circle rotates, Fig. 1a (or equivalently the primary electron beam precesses, Fig. 1b). All reflections intersect the Ewald sphere twice per precession cycle and (at least) partially integrated reflection intensities are recorded. Note that the locations of electron diffraction reflections on properly de-scanned (single-crystal) PED patterns, Fig. 1c, are the same as those of their (non-integrated) counterparts on conventional (single-crystal), stationary-beam selected area electron diffraction (SAED) patterns, Figs. 2a and 3a.

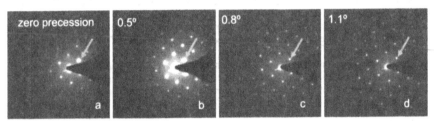

Fig. 2: Experimental diffraction patterns from a silicon crystal, approximately 60 nm thickness, orientation close to [110], 200 kV. **(a)** SAED pattern (zero precession), **(b)**, **(c)** and **(d)** PED patterns from the same area with increasing precession angle. Note that while the intensity of the $(1\bar{1}\bar{1})$ reflection, marked by arrows, is much higher than that of its Friedel pair $(\bar{1}11)$ and that of the other two symmetry equivalent $\pm(1\bar{1}1)$ reflections in the SAED (a), the intensities of all four {111} reflections are very similar for the PED patterns (b), (c) and (d).

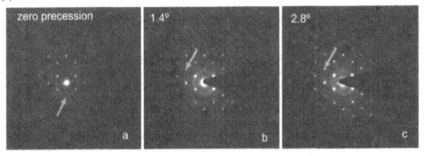

Fig. 3: Experimental diffraction patterns from an approximately 6 nm thin section of a wedge shaped silicon crystal in [110] orientation, 200 kV. **(a)** SAED pattern (zero precession); **(b)** and **(c)** PED patterns from the same sample area with increasing precession angle. While the (barely visible) kinematically forbidden (002) reflection is marked by an arrow in (a), the arrows in (b) and (c) mark the kinematically forbidden $(\bar{2}2\bar{2})$.

Because individual reflections are excited more or less sequentially while a PED pattern is building up, secondary dynamical scattering between simultaneously excited reflections is significantly reduced. For the same crystal thickness, the electron scattering conditions approach in the PED mode the assumptions of the two-beam dynamical diffraction theory much more closely than in the SAED mode. Because PED reflection intensities are also (at least partially) integrated, the Blackman correction for primary extinction effects [16] is (at least in principle) applicable for inorganic crystal with thicknesses of a few tens of nanometers [2,3,17].

Dynamical systematic row scattering is, however, not suppressed by the PED geometry since such rows tend to be excited at once [2,3,17]. Some "residual" secondary scattering also contributes to the intensities of kinematically forbidden reflections, but can be reduced by increasing the precession angle. The kinematically forbidden {222} reflections are, for example, visible in the experimental

PED patterns of Figs. 2b-d and 3b. The kinematically forbidden reflections can, however, be differentiated from the other reflections by their characteristic intensity reduction with increasing precession angle [2,3] Also, at least in theory, the kinematically forbidden reflections possess a different shape. The thinner the crystal and the higher the precession angle, the more the reflection intensity distribution in an experimental PED pattern will approach the kinematic limit of the two-beam dynamical theory. The integration of the reflection intensities will also be more complete with higher precession angles.

The radius of the Laue circle depends on the precession angle and can be calculated from the relation $R = \lambda \sin \varepsilon$, where ε is the precession angle, i.e. the half angle of the hollow illumination cone of the precessing primary electron beam. An experimental PED pattern may extend in reciprocal space approximately to twice this radius. For 200 kV electrons and a precession angle of 2.8°, one obtains about 20 nm^{-1} for this extension, corresponding to a direct space resolution of 0.5 Å. This increased resolution with respect to conventional SAED patterns may be understood as resulting from an "effective flattening of the Ewald sphere" and is illustrated by the experimental diffraction patterns of Figs. 2b-d and 3b,c. The integrated reflection intensities in PED patterns are modified by the prevailing Lorentz factors, which depend on the precession angle, crystal structure and thickness, as well as on the type of the crystalline sample [14,15].

Figure 2 illustrates that the recording of a PED pattern is experimentally less demanding than the recording of a SAED pattern since characteristic "zone axis diffraction patterns" can be obtained for crystal that are slightly mis-oriented. This is because the precession movement and proper de-scanning lead to an (at least partial) integration of the reflection intensities over the excitation error.

STRUCTURAL FINGERPRINTING FROM PED SPOTS OF INDIVIDUAL NANOCRYSTALS

Although the silicon crystal of Fig. 2 is approximately 60 nm thick, the reflection intensities in the PED patterns may still be used for structural fingerprinting in either the kinematic ($I_{hkl} \sim F_{hkl}^2$, relation 5) or the asymptotic ($I_{hkl} \sim F_{hkl}$) limit of the two-bean dynamical theory [15]. The reflection intensities of the approximately 6 nm thin silicon crystal of Fig. 3, on the other hand, can be treated kinematically as dynamical diffraction effects are negligible. This can be safely inferred from the very low intensity of the kinematically forbidden {002} and {222} reflections in Fig. 3a. One must, however, take the Lorentz factor into account for the PED patterns of the 6 nm thin silicon crystal. This requirement is demonstrated by the clear visibility of the kinematically forbidden {222} reflections in Figs. 3b and 3c.

Because the reflection intensities are (at least partially) integrated, the Laue class symmetry in the PED patterns can be made an additional component of structural fingerprinting. Due to the small curvature of the Ewald sphere, the possible Laue classes of the reflections in the zero order Laue zone all contain a two-fold rotation axis as the projection of a center of symmetry onto a plane. One can, therefore, only distinguish between those six 2D point groups that contain a two-fold rotation axis, i.e. 2, 2mm, 4, 4mm, 6 and 6mm. The other four 2D point groups that do not contain such an axis, i.e. 1, m, 3, and 3m, are possible for reflections in higher order Laue zones, since {h k l+1} and {-h -k -l+1} reflections are not Friedel pairs. Note also that the Laue class symmetry of PED patterns that were recorded with large precession angles is rather insensitive to the exact crystal orientation and dynamical diffraction effects. It is, therefore, a valuable characteristic that can be employed for structural fingerprinting from PED patterns.

Since kinematically forbidden reflections can be identified from a series of PED patterns with increasing precession angle [2,3], space group information may be inferred from them [20]. Reflections from higher order Laue zones in large precession angle PED patterns, e.g. Fig. 1a, are particularly useful for this purpose. The space group information determination can be utilized as a third component of structural fingerprinting of individual nanocrystals in the TEM [2,3]. This is all in addition to traditional structural fingerprinting of individual single-crystals from SAED patterns were only the projected reciprocal lattice geometry, i.e. the two shortest reciprocal spacings and the interplanar angle have been utilized [1].

AUTOMATED CRYSTALLITE ORIENTATION & STRUCTURE MAPPING FROM SCANNING PED SPOT PATTERNS AND HIGH-RESOLUTION TEM IMAGES

The automated mapping procedure comprises the collection of PED patterns (by either the internal slow-scan charged coupled device (CCD) camera or by a much faster external digital camera) while the sample area of interest is scanned and precessed with a tens of nanometers sized primary electron beam. The data collection step is followed by automatic data processing, Fig. 4 left. The orientation and structure identification is performed by the matching of each experimental PED pattern to a series of templates, which are pre-calculated theoretical diffraction patterns for all symmetry equivalent orientations. A typical processing rate for the matching of PED patterns of cubic nanocrystals is approximately 100 patterns per second. The degree of matching between the experimental PED patterns (Fig. 4 right, section a) and the calculated templates (Fig. 4 right, section b) is given by a correlation index [7].

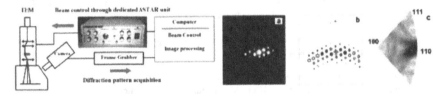

Fig. 4 from ref. [7], **Left:** Sketch of the experimental set up of the ASTAR system (from NanoMEGAS SPRL*) for automated crystallite orientation and structure mapping. **Right:** Sketch of the PED pattern matching procedure: (**a**) an experimental PED pattern from a crystallite that was acquired by an external digital camera is compared to (**b**) simulated templates until the best match is found; the correlation index map (**c**) reflects the degree of matching. The dot in (c) shows the most probable orientation of the cubic crystallite that led to the PED pattern of (a).

The spatial resolution of the automatically generated crystallite orientation and structure maps depends on the primary electron beam diameter and the scanning-precessing-probe step-size. The spatial resolution of these maps can, therefore, be made superior to the maps that can be recorded by means of the electron backscattering (Kikuchi) diffraction technique (EBSD) in a scanning electron microscope (SEM). In addition, electron spot diffraction in a TEM is much less sensitive to the plastic deformation state and possible surface contaminations of nanocrystals that EBSD in a SEM.
In our study of a mixture of magnetite and maghemite nanocrystals, the scanning-precessing-probe step-size was approximately 30 nm, while the primary electron beam had a diameter of about 20 nm [4]. We recorded thousands of PED patterns with a precession angle of approximately 0.3° at the prototype site of the ASTAR* system in Grenoble [7]. The experimental data was used to generate the crystallite orientation and structure maps of Fig. 5. A comparison between the correlation index map, Fig. 5b, and the crystallite structure map for magnetite, Fig. 5d, shows that the majority of the nanocrystals possess the magnetite structure. The same conclusion was reached from our earlier structural fingerprinting studies from HRTEM images of individual crystallites of this mixture [8-10].

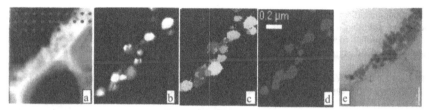

Fig. 5: Iron-oxide nanocrystals in a TEM; **(a)** to **(d)** automated crystallite orientation & structure mapping; **(e)** conventional (parallel-illumination) bright-field image of these crystallites (for comparison purposes only). **(a)** Map of the fluctuations of the primary electron beam intensity in the PED patterns, which is comparable to a bright-field scanning TEM image. The array of spots represents the scanning and precessing of the electron probe in steps (that are not to the actual scale). **(b)** Correlation index map; **(c)** Crystallite orientation map; **(d)** Crystallite structure map of magnetite.

An extension of this automated analysis approach to Fourier transforms of parallel-illumination HRTEM images of nanocrystals is illustrated in Fig. 6 [11]. The experimental data for this application was recorded on the internal slow-scan CCD cameras of objective-lens aberration-corrected TEMs. The projected reciprocal lattice geometry of a nanocrystal is revealed by the modulus part of the Fourier transform (which is typically displayed in conjunction with HRTEM images and considered to be a "digital diffractogram" of the former). Just as ordinary diffraction patterns, this part of the Fourier transform of HRTEM images can be utilized for the first step of advanced structural fingerprinting and may subsequently be augmented with information on the projected symmetry and individual structure factors to result in a sufficiently discriminatory fingerprinting procedure [1,8-10].

While nanocrystals with relatively large unit cells, i.e. magnetite and maghemite (a ≈ 0.84 nm), can be structurally fingerprinted in the HRTEM imaging mode with modern analytical TEMs or older "dedicated HRTEMs" (with typical point-to-point resolutions of 0.24 to 0.19 nm, e.g. refs. [8-10]), many other inorganic nanocrystals require the utilization of aberration-corrected microscopy for the recording of the data. Figure 7 illustrates this necessity on gold nanocrystals (a ≈ 0.41 nm). Table I gives the crystallographic indices of lattice fringes and zone axes that can (at least in principle) be derived from HRTEM images as a function of the point-to-point resolution of a TEM. A hypothetical cubic AB-compound with halite structure and small lattice constant was chosen as example. While the top two rows of this table refer to dedicated HRTEMs, the bottom two rows refer to aberration-corrected TEMs [2].

Table I: Relationship between point-to-point resolution of a TEM and the principle visibility of lattice fringes and zone axes within one stereographic triangle [001]-[011]-[111] for a hypothetical cubic AB-compound with 0.425 nm lattice constant and space group $Fm\overline{3}m$, (i.e. the halite structural prototype to which PbSe belongs).

Point-to point resolution [nm]	Number and type of visible lattice fringes (net-plane families)	Number and type of visible zone axes (lattice fringe crossings)
0.2	**2**, i.e. {111}, {200}	**2**, i.e. [001], [011]
0.15	**3**, i.e. {111}, {200}, {220}	2^2, i.e. [001], [011], [111], [112]
0.1	**4**, i.e. {111}, {200}, {220}, {311}	2^3, i.e. [001], [011], [111], [112], [013], [114], [125], [233]
≈ 0.05	≈ **18**, i.e. {111}, {200}, {220}, {311}, {331}, {420}, {422}, {511}, {531}, {442}, {620}, {622}, {551}, {711}, {640}, {642}, {731}, {820}	> 2^5, e.g. [001], [011], [111], [012], [112], [013], [122], [113], [114], [123], [015], [133], [125], [233], [116], [134], [035], ...

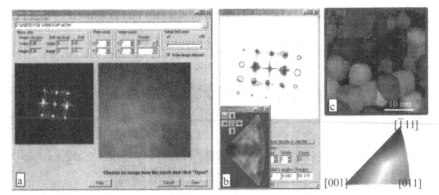

Fig. 6: Processing of HRTEM images of PbSe (a ≈ 0.61 nm) nanocrystals with software that complements the ASTAR* system. **(a)** Overview HRTEM image for selection of nanocrystals for calculation of fast Fourier transform (FFT). **(b)** Simulated template that matches a particular FFT best overlaid (upper part of panel) and location of that template in the stereographic triangle for point groups *432*, $m\overline{3}$, and *23* (as inset). **(c)** Color-coded crystallite orientation map of PbSe nanocrystals with stereographic-triangle color-key for point groups $\overline{4}3m$ and $m\overline{3}m$.

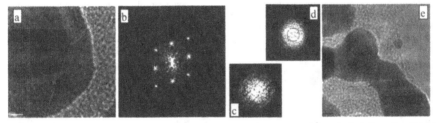

Fig. 7: Increase in "crystallographic resolution" as a function of correction of the spherical aberration (C_s) of the FEI Super Twin objective lens on the example of nanocrystalline gold. While the C_s is on the order of magnitude a few tens to a few hundreds of μm in the HRTEM image shown as **(a)** since it was recorded in an FEI Titan 80-300 (at 300 kV), it is approximately 1.2 mm in the "classical" HRTEM image **(e)** that was recorded in an analytical FEI Tecnai G^2 F20 (at 200 kV.) Note that both TEMs have essentially the same objective lens, but C_s can be significantly reduced by the fine tuning of additional electron optical components in the Titan microscope. (Note the pronounced reduction in contrast delocalization. There is a slight increase of resolution due to the increase in the acceleration voltage.) As revealed in the digital diffractogram **(b)** of an approximately 14 nm² wide section of (a), 16 data points can be utilized for structural fingerprinting on the basis of the projected reciprocal lattice geometry even for this small lattice constant material. Digital diffractograms from other HRTEM images that were calculated for approximately 4,000 nm² wide sections of images taken from ensembles of nanocrystalline gold in the Titan **(c)** and the Tecnai **(d)** microscope demonstrate that much smaller spacings can be reliably resolved with the former.

Besides its obvious utility for a wide range of materials science applications, automated crystallite orientation & structure mapping in a TEM from either single crystal PED patterns or HRTEM images complements both structural fingerprinting from individual nanocrystals [1-11] and structural electron crystallography [12-15]. This is because the crystallite orientation maps may be used for the selection of individual nanocrystals that are oriented close to major zone axes.

STRUCTURAL FINGERPRINTING FROM POWDER PED PATTERNS OF ENSEMBLES OF NANOCRYSTALS

A precessing primary electron beam that is up to hundreds of nanometers wide may also benefit structural fingerprinting from ensembles of nanocrystals, Figs. 8 and 9.

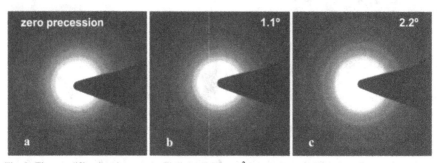

Fig. 8: Electron diffraction ring patterns from the same (μm^2-sized) area of a fine-grained crystal-powder of Ni-doped cassiterite (SnO_2) nanocrystals (deposited from toluene). **(a)** Powder SAED pattern, zero precession; **(b)** and **(c)** powder PED patterns with increasing precession angle. These nanocrystals possess an average size of approximately 3 nm, are paramagnetic, stable as colloids in non-polar solvents, and become ferromagnetic at room temperature when capped by trioctylphosphine oxide, spin coated into thin films on fused silica substrates, and calcined [21]. (The apparently high inelastic background in these patterns and those shown below in Fig. 8 is likely to be due to the prior suspension of the nanocrystals for storage in toluene, which we assume to stick partly to the amorphous carbon support films that cover the Cu grids of both TEM samples.)

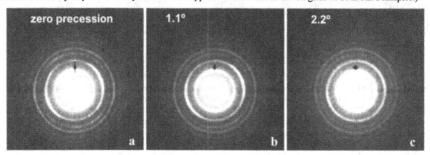

Fig. 9: Electron diffraction ring patterns from the same (μm^2-sized area) of Co-doped rutile (TiO_2) nanocrystals (deposited from toluene). **(a)** Powder SAED pattern, zero precession; **(b)** and **(c)** powder PED patterns with increasing precession angle. These nanocrystals are elongated in the crystallographic c-direction, possess an average size of 4 to 6 nm, and are ferromagnetic at room temperature after processing into thin films [22]. (The conspicuous dark specks and the rather faint horizontal stripes at about the middle of the patterns are quite unimportant artifacts of the data recording process.)

Just as in PED patterns from single crystals, dynamical multiple-beam diffraction effects are suppressed in "powder PED patterns". In combination with electron-energy filtering to reliably remove the inelastically scattered background [23], this may result in the high accuracy recovery of structure factor moduli from the intensities of the Debye-Scherrer rings. Because more nanocrystals contribute to these reflection rings in the precession powder mode, the elastically scattered intensities in these patterns are also based on better nanocrystal statistics. Consequently, in comparison to standard powder SAED patterns from the same sample area, the Debye-Scherrer rings are more uniform when the precession electron diffraction geometry is utilized, Figs. 8 and 9. While the influence of a weak fiber texture on the reflection ring intensities is not negligible in Figs. 9a and 9b and results in inhomogeneities in the Debye-Scherrer rings, powder PED patterns with a sufficiently large precession angle, Fig. 9c, possess more homogeneous rings and are, therefore, more valuable for structural fingerprinting in the TEM.

SUITABLE CRYSTALLOGRAPHIC DATABASES

Traditional structural fingerprinting [1] typically relies on the Powder Diffraction File (PDF-2, the number after the acronym signifies that it is the second format release) [24] and the Crystal Data database of the National Institute of Standards and Technology (NIST) [25]. None of these databases, however, contains sufficient information to calculate structure factors (using relation 2), electron reflection intensities, or 2D Laue class symmetries. There are also no means to obtain the kinematically forbidden reflections from these databases.

The Inorganic Crystal Structure Database [26], NIST's Standard Reference Data Bases 83 and 84 [27], the Pearson's Crystal Data [28] database, and the PDF-4 [24], on the other hand, provide the necessary information for inorganic crystals. As an alternative to these commercial databases, one may now use comprehensive open-access databases. Some 100,000 comprehensive datasets are available at the combined web sites of the Crystallography Open Database [29,30], its mainly inorganic and educational subset [31], the American Mineralogist Crystal Structure Database [32], and the Linus Pauling File [33]. More information on crystallographic databases in general (and the multitude of smaller open-access crystallographic databases) is available over the Uniform Resource Locators given as refs. [34,35].

SUMMARY AND CONCLUSIONS

Strategies for the structurally identification of nanocrystals from precession electron diffraction patterns were outlined. While a single-crystal PED pattern is used for the structural identification of an individual nanocrystal, ensembles of nanocrystals are fingerprinted structurally from powder PED patterns. Highly reliable crystal orientation & structure maps can be obtained from automatically collected/processed single-crystal scanning-PED patterns and semi-automatically processed parallel illumination HRTEM images. Open-access crystallographic databases support our structural fingerprinting strategies because they are comprehensive and contain all of the necessary crystallographic information.

ACKNOWLEDGMENTS

This research was supported by awards from the Oregon Nanoscience and Microtechnologies Institute. Additional support from Portland State University's Venture Development Fund is acknowledged. Prof. Daniel R. Gamelin of the University of Washington at Seattle and Dr. Klaus H. Pecher of the Pacific Northwest National Laboratory are thanked for the cassiterite, rutile, and iron-oxide samples. Dr. Kurt Langworthy of the University of Oregon's Center for Advanced Materials Characterization in Oregon is thanked for his assistance in operating the aberration-corrected FEI Titan 80-300 microscope. Prof. Marie Cheynet of the Institut National Polytechnique de Grenoble is thanked for the HRTEM image of Fig. 6 (which was taken with an aberration-corrected FEI Titan 80-300 microscope in Grenoble) from PbSe nanocrystals that were produced by Dr. Odile Robbe of the Université de Lille.

REFERENCES

[1] Moeck, P.; Fraundorf, P.: *Zeits. Kristallogr.* **222** (2007) 634-645, expanded version at: **arXiv:0706.2021**

[2] Moeck, P.; Rouvimov, S.: *in: Nano Particle Drug Delivery Systems: II Formulation and Characterization*, Y. Pathak and D. Thassu (editors), Informa Health Care, New York, 2009, 268-311.

[3] Moeck, P.; Rouvimov, S.: *Zeits. Kristallogr.* (2009) *in press*

[4] Rouvimov, S.; Rauch, E. F.; Moeck, P.; Nicolopoulos, S.: *Proc. NSTI 2009*, Houston, Texas, *in press*

[5] Moeck, P.; Rouvimov, S.: *Proc. NSTI 2009*, Houston, Texas, *in press*

[6] Moeck, P.; Rouvimov, S.; Nicolopoulos, S.: *Proc. NSTI 2009*, Houston, Texas, *in press*

[7] Rauch, E. F.; Véron, M.; Portillo, J.; Bultreys, D.; Maniette Y.; Nicolopoulos, S.: *Microscopy and Analysis*, Issue 93, November 2008, S5-S8.

[8] Bjorge, R.: *MSc thesis*, Portland State University, May 9, 2007; Journal of Dissertation Vol. 1 (2007), **open access:** http://www.scientificjournals.org/journals2007/j_of_dissertation.htm

[9] Moeck, P.; Bjorge, R.: *in:* Quantitative Electron Microscopy for Materials Science, Eds. E. Snoeck, R. Dunin-Borkowski, J. Verbeeck, and U. Dahmen, *Mater. Res. Soc. Symp. Proc.* Vol. **1026E** (2007), paper 1026-C17-10.

[10] Moeck, P.; Bjorge, R.; Mandell, E.; Fraundorf, P.: *Proc. NSTI-Nanotech* Vol. **4** (2007) 93-96, (www.nsti.org, ISBN 1-4200637-6-6).

[11] Rauch, E. F.; Rouvimov, S.; Nicolopoulos, S.; Moeck, P.: *Proc. Microscopy & Microanalysis* 2009, Richmond, Virginia, *in press*

[12] Zou, X. D.; Hovmöller, S.: *Acta Cryst. A* **64** (2008) 149-160; **open-access:** http://journals.iucr.org/a/issues/2008/01/00/issconts.html

[13] Dorset, D. L.: *Structural Electron Crystallography*, Plenum Press, New York and London, 1995.

[14] Vainshtein, B. K.; Zvyagin, B. B.: *in: International Tables for Crystallography*, Vol. **B**, Reciprocal space, Ed. U. Shmueli, 2nd edition, Kluver Academic Publ., Dordrecht, 2001, pp. 306-320.

[15] Vainshtein, B. K.: *Structure Analysis by Electron Diffraction*, Pergamon Press Ltd., Oxford, 1964.

[16] Blackman, M.: *Proc. Royal Society (London)* A **173** (1939) 68-82.

[17] Vincent, R.; Midgley, P.: *Ultramicroscopy* **53** (1994) 271-282.

[18] Avilov, A.; Kuligin, K.; Nicolopoulos, S.; Nickolskiy, M.; Boulahya, K.; Portillo, J.; Lepeshov, G.; Sobolev, B.; Collette, J. P.; Martin, N.; Robins, A. C.; Fischione, P.: *Ultramicroscopy* **107** (2007) 431-444.

[19] Klechkovskaya, V. V.; Imamov, R. M.: *Crystallography Reports* **46** (2001) 534-549.

[20] Morniroli, J. P.; Redjaïmia, A.: *J. Microsc.* **227** (2007) 157-171.

[21] Archer, P. I.; Radovanovic, P. V.; Heald, S. M.; Gamelin, D. R.: *J. Am. Chem. Soc.* **127** (2005) 14479-14487.

[22] Bryan, J. D.; Heald, S. M.; Chambers, S. A.; Gamelin, D. R.: *J. Am. Chem. Soc.,* **126** (2004) 11640-11647.

[23] Cowley, J. M.: *Progress in Materials Science*, Vol. **13**, 267-321, Eds. B. Chalmers, and W. Hume-Rothery, Pergamon Press, Oxford, 1967.

[24] Faber, J.; Fawcett, T.: *Acta Cryst.* **B 58** (2002) 325-332, http://www.icdd.com

[25] Mighell, A. D.; Karen, V. L.: *J. Res. Natl. Inst. Stand. Technol.* **101** (1996) 273-280; *NIST Standard Reference Database 3*, http://www.nist.gov/srd/nist3.htm

[26] http://www.fiz-karlsruhe.de/icsd.html, about 3,600 entry on-line demo version freely accessible at: http://icsdweb.fiz-karlsruhe.de/

[27] http://www.nist.gov/srd/nist83.htm *and* http://www.nist.gov/srd/nist84.htm (also free download of an about 3,200 inorganics entry demo version of ref. [27]

[28] http://www.crystalimpact.com/pcd/Default.htm, about 2,600 entry demo version for free download at: http://www.crystalimpact.com/pcd/download.htm

[29] http://www.crystallography.net *mirrored at:* http://cod.ibt.lt (in Lithuania), http://cod.ensicaen.fr/ (in France) *and* http://nanocrystallography.org (in Oregon, USA), *also accessible under a different search surface at:* http://fireball.phys.wvu.edu/cod/ (in West Virginia, USA), some 75,000 data sets

[30] Gražulis, S.; Chateigner, D.; Downs, R. T.; Yokochi, A. F. T.; Quirós, M.; Lutterotti, L.; Manakova, E.; Butkus, J.; Moeck, P.; Le Bail, A.: *J. Appl. Cryst., submitted*

[31] http://nanocrystallography.research.pdx.edu/CIF-searchable/cod.php, data on some 20,000 crystals

[32] http://rruff.geo.arizona.edu/AMS/amcsd.php, data on some 10,000 minerals

[33] http://crystdb.nims.go.jp, data on some 30,000 metals and alloys

[34] http://en.wikipedia.org/wiki/Crystallographic_database

[35] http://nanocrystallography.research.pdx.edu/index.py/group_links

* NanoMEGAS SPRL (www.nanomegas.com) offers such devices and Portland State University's "Laboratory for Structural Fingerprinting and Electron Crystallography" (run by Prof. Peter Moeck) serves as the first demonstration site of this company in the Americas. A second generation precession electron diffraction device "Spinning Star" and the ASTAR software are installed there on an analytical FEI Tecnai G^2 F20 ST transmission electron microscope and can be demonstrated on request.

Image Deconvolution-An Effective Tool of Crystal Structure and Defect Determination in High-Resolution Electron Microscopy

Fanghua Li and Chunyan Tang*

Beijing National Laboratory for Condensed Matter Physics, Institute of Physics, Chinese Academy of Sciences, P.O. Box 603, Beijing 100190, China
* Current address: Biology Department, Brookhaven National Laboratory, NY 11973, U.S.A.

ABSTRACT

Image deconvolution is introduced as an effective tool to enhance the determination of crystal structures and defects in high-resolution electron microscopy. The essence is to transform a single image that does not intuitively represent the examined crystal structure into the structure image. The principle and method of image deconvolution together with the related image contrast theory, the pseudo weak phase object approximation (pseudo WPOA), are briefly described. The method has been applied to different types of dislocations, twin boundaries, stacking faults, and one-dimensional incommensurate modulated structures. Results on the semiconducting epilayers $Si_{0.76}Ge_{0.24}$/Si and $3C$-SiC/Si are given in some detail. The results on other compounds including AlSb/GaAs, GaN, $Y_{0.6}Na_{0.4}Ba_2Cu_{2.7}Zn_{0.3}O_{7-\delta}$, $Ca_{0.28}Ba_{0.72}Nb_2O_6$ and $Bi_{2.31}Sr_{1.69}CuO_{6+\delta}$ are briefly summarized. It is also shown how to recognize atoms of Si from C based on the pseudo WPOA, when the defect structures in SiC was determined at the atomic level with a 200 kV LaB_6 microscope.

INTRODUCTION

The development of high-resolution electron microscopy (HREM) has resulted in a powerful technique that allows for the direct observation of projected crystal structures [1, 2]. Figure 1 gives a schematic diagram of the image formation process in HREM. Passing through an object, the electron wave is modulated to form the exit wave $q(r)$ on the bottom surface of the object. The exit wave carries the information of the object structure and becomes the object wave for the objective lens. The diffracted and image waves are then formed on the back focal plane and image plane of objective lens, respectively. The diffracted wave function $Q(H)$ is the Fourier transform (FT) of the object wave, and the image wave $\psi(r)$ is the inverse FT of the diffracted wave modulated by the transfer function of the objective lens. The transfer function depends on various electron-optical parameters such as the defocus, spherical and chromatic aberration of the lens, beam divergence, etc. Generally, only images taken near the Scherzer defocus (for short hereafter written as Scherzer focus) [3] represent the projected object structure faithfully when the sample is sufficiently thin, while an image taken with an arbitrary defocus does not directly reflect the structure. In addition, due to the typically limited microscope resolution, atoms may not be resolved individually even in images that were taken at the Scherzer focus.

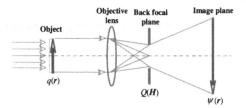

Figure 1. Schematic diagram showing the image formation process. $q(r)$, $Q(H)$ and $\psi(r)$ are exit wave, diffracted wave and image wave, respectively.

Different methods of posterior image processing were developed to retrieve the structures from experimental images that did not directly represent the structures. For instance, the through focus series technique has been developed and applied to obtain the exit wave on the bottom surface of the object from a series of images taken under different defocus condition [4, 5], and the electron crystallographic image processing technique to derive the structure from a single image or from an image together with the corresponding electron diffraction pattern [6, 7]. The resolution of retrieved structure images is usually higher than the point resolution of the electron microscope and close to its information limit. Furthermore, it will reach the diffraction resolution limit when electron diffraction data are utilized to enhance the image information.

Two sets of electron crystallographic image processing techniques have been developed in the Institute of Physics, Chinese Academy of Sciences [7]. One of them aims at the *ab initio* crystal structure determination by means of image deconvolution followed by structure factor phase extension from an image and the corresponding electron diffraction pattern. The other aims at deriving the local and long range aperiodic structures in crystals such as line and planar defects, as well as incommensurately modulated structures from a single image via deconvolution processing only. This paper concentrates on the image deconvolution for crystals with defects and describes the respective principle, method and applications.

PRINCIPLE AND METHOD

Principle of image deconvolution

The weak phase object approximation (WPOA) clearly and simply interprets the one-to-one correspondence between the projected crystal structure and the structure image taken at or close to the Scherzer focus. Under the WPOA, the image intensity is expressed as

$$I(r) = 1 + 2\sigma\varphi(r) * \mathcal{F}^{-1}[T(H)] \qquad (1)$$

where $\sigma = \pi/\lambda U$, λ denotes the electron wavelength and U the accelerating voltage, $\varphi(r)$

represents the projected electrostatic potential distribution function, $T(H)$ is the contrast transfer function (CTF), $*$ and $\mathcal{F}^{-1}$ are operators of convolution and inverse FT, respectively. The CTF depends on various electron-optical parameters such as the spherical aberration coefficient of the objective lens, the defocus amount, the defocus spread due to the chromatic aberration, and others. Figure 2 shows two CTF curves corresponding to 200 kV microscopes equipped with a field emission gun (FEG) and a LaB_6 filament, respectively. The oscillation of the CTF curves with a change of defocus leads to a modulation of both phases and amplitudes of diffracted waves so that an arbitrary image may not represent the crystal structure. In a rather narrow defocus range, the CTF curve has a more or less flat region with $T(H) \approx -1$, and the image intensity is approximately expressed as

$$I(r) = 1 - 2\sigma\varphi(r) \qquad (2)$$

The contrast of such an image, so called "Scherzer focus" image, is linear to the projected potential. The image deconvolution aims at removing the CTF modulation in order to retrieve a genuine projection of the structure from a single image and enhancing the image resolution up to the information limit of the microscope.

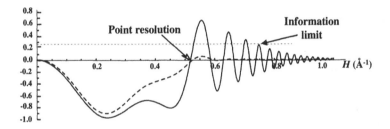

Figure 2. CTF curves for 200 kV field-emission gun (solid) and LaB_6 filament (dotted) electron microscopes.

Because most crystals observed in electron microscopes are thicker than a weak phase object, a modified image intensity expression was derived in 1985 to meet the practical crystal thickness range in HREM. The corresponding image contrast theory was named the pseudo weak phase object approximation (pseudo WPOA) [8]. The new formula for the Scherzer focus image is very similar to equation (2), but the function $\varphi(r)$ is replaced by $\varphi'(r)$ which represents the potential of an artificial crystal which has the same structure as the examined real structure, but the constituent atoms are different such that the heavy atoms in the artificial crystal are lighter than those in the real crystal, and vise versa. Hereafter, $\varphi'(r)$ is named the pseudo electrostatic projected potential distribution function or in short pseudo potential. Approximately, we have

$$I(r) = 1 + 2\sigma\varphi'(r) * \mathcal{F}^{-1}[T(H)] \tag{3}$$

The FT of the image intensity is expressed as

$$i(H) = 1 + 2\sigma F'(H)T(H) \tag{4}$$

where $i(H)$ and $F'(H)$ denote the FTs of $I(r)$ and $\varphi'(r)$, respectively. $F'(H)$ is the structure factor of the artificial crystal mentioned above and named pseudo structure factor. Formula (4) implies that it is reasonable to apply various analysis methods developed in diffraction crystallography to $F'(H)$. Omitting the transmitted beam in equation (4) yields

$$F'(H) = \frac{i(H)}{2\sigma T(H)} \tag{5}$$

Equation (5), employing reciprocal space representations, demonstrates that it is straightforward to deal with the modulation due to the CTF in order to obtain the pseudo structure factor $F'(H)$, when $T(H)$ is known. Then the image deconvolution can be obtained by inversely Fourier transforming $F'(H)$ to yield the pseudo potential $\varphi'(r)$.

Equation (4) is valid when the thickness of the observed crystal is below a critical value, which depends on the electron wavelength and the weights of the constituent atoms in the examined crystals [8]. The critical crystal thickness is generally less than 10 nm for 200 and 300 kV microscopes, but may be equal or larger than 10 nm for crystals containing rather light atoms.

Method of deconvolution for crystal defects

This method consists of two components: defocus determination and deconvolution. In addition, the dynamical scattering effect correction [9] can be utilized as a supplement to the recognition of different kinds of atoms in the perfect crystalline image regions, before moving on to the defect determination. Figure 3 gives the flow chart of the image deconvolution procedure.

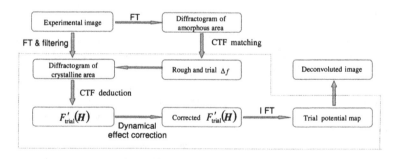

Figure 3. Flow chart of image deconvolution for crystals with defects.

All parameters included in $T(H)$ are usually known except the defocus amount $\Box f$. In case of perfect crystal structure determination, the method is based on the principle of maximum entropy and has been well studied and proved to be very effective to determine the defocus amount in the process of the image deconvolution [10-12]. In the present case a rough defocus amount is determined first either from an adjacent perfect crystalline image region or from an amorphous region nearby the examined crystalline area. In the latter case, the Thon diffractogram [13] is utilized. Then the defocus refinement is performed in the process of image deconvolution. For this purpose several trial defocus values close to the roughly determined one are assigned and small focus steps, for instance 0.5 nm, are used. For each trial defocus value the modulation due to the CTF is removed from the diffractogram of the image containing the examined defect. An artificial unit cell with the examined defect area at the center is constructed, and for each trial defocus a set of $F'_{\text{trial}}(H)$ is calculated from equation (5).

To reduce the dynamical scattering effect, all $F'_{\text{trial}}(H)$ sets are corrected by forcing the integrated amplitudes of reflections to be equal to the amplitudes of corresponding structure factors for the perfect crystal $F(H)$ [9]. The corrected $F'_{\text{trial}}(H)$ for the i-th pixel is

$$\left| F'_{\text{trial}}(H) \right|_i^{\text{corr}} = K_H \left| F'_{\text{trial}}(H) \right|_i \tag{6}$$

and

$$K_H = \frac{\left| F(H) \right|}{\sum_i \left| F'_{\text{trial}}(H) \right|_i} \tag{7}$$

where $F(H)$ is the structure factor for the perfect crystal, K_H is the correction coefficient that is constant for all pixels of the same reflection and different for different reflections. Thus, inversely Fourier transforming all corrected $F'_{\text{trial}}(H)$ sets yields the trial potential maps of the crystal with the defects. Finally, the best map is selected as the correctly deconvoluted image from among all of these trial maps. In this map, atoms in both the perfect and defect regions should be resolved most clearly.

APPLICATIONS

The deconvolution processing has been applied to images of various compounds. Different defect structures have been successfully retrieved, for instance, the core structures of Lomer dislocation and 60° dislocation complex in $Si_{0.76}Ge_{0.24}/Si$ [14, 15], the twin boundary structure and 30° partial dislocation core structure in $3C$-SiC [16], the misfit dislocation core structures at AlSb/GaAs interface [17], the stacking fault in GaN [18], the twinning dislocation in $Y_{0.6}Na_{0.4}Ba_2Cu_{2.7}Zn_{0.3}O_{7-\delta}$ [19], and the one-dimensional incommensurate modulated structures of $Ca_{0.28}Ba_{0.72}Nb_2O_6$ and $Bi_{2.31}Sr_{1.69}CuO_{6+\delta}$ [20, 21]. The original images were taken with 200 kV electron microscopes, either equipped with a LaB_6 filament or a field-emission gun. In order to determine the defect structures at the atomic level in SiC and AlSb, atoms of Si were distinguished from C and Al from Sb in the deconvoluted images employing the pseudo WPOA, respectively. In the following, only the research results on $Si_{0.76}Ge_{0.24}$/Si and $3C$-SiC are illustrated. Except for cases where the type of the microscope is explicitly specified, all images were taken with a 200 kV LaB_6 microscope of point resolution 0.2 nm. The key points and detailed procedure of deconvolution processing will be illustrated in the first example, a Lomer dislocation in $Si_{0.76}Ge_{0.24}$/Si.

Dislocation core structures in epilayer SiGe/Si

(1) *Lomer dislocation* [14]

The crystal structure of $Si_{0.76}Ge_{0.24}$/Si is isomorphic to that of Si. In the [110] projected structure of $Si_{0.76}Ge_{0.24}$/Si, the distance between two adjacent atoms is about 0.14 nm, which is close to the information limit of 200 kV FEG electron microscopes. Several images were taken under different defocus conditions. The FT was performed for each image to obtain the diffractogram. To reveal the atoms individually by image deconvolution, the structure information with all spatial frequencies up to $(0.14 \text{ nm})^{-1}$ must be contained in the original image. This requires that the reflection 004 must appear in the diffractogram. In addition, since there are four independent reflections with the indexes 111, 220, 113 and 004 up to the spatial frequency $(0.14 \text{ nm})^{-1}$ for perfect crystals, all four independent reflections must appear in the diffractogram in order to avoid a large loss of information. The image with the best quality was selected for the following image deconvolution processing.

Figure 4(a) is an area from a [110] projection high-resolution image taken with a JEOL JEM-2010F electron microscope. A dislocation is seen at the center. Although it can be recognized from the image that the dislocation is of the Lomer type, the atomic configuration of the dislocation core is not revealed directly. Several models have been proposed for the Lomer dislocation. Bourret, Desseaux and Renault proposed two symmetric and two asymmetric core models in Si and Ge [22]. McGibbon, Pennycook and Angelo claimed to have observed the Hornstra-like and an unexpected core structure in the polar compounds CdTe/GaAs [23]. In case of SiGe/Si, no sufficient evidence has been found before to support any of the models mentioned above or to propose a new one. Figure 4(b) is the diffractogram obtained from a circular area with a diameter of about 17.6 nm, with the dislocation at the center. All reflections within the spatial frequency range for resolving the two adjacent atoms appear with rather dominant intensity. This implies that no reflection falls in the vicinity of zero crossings of CTF. The FT of the image area including both the amorphous and crystalline regions near the dislocation yields the picture showing the superposition of Thon diffractogram and diffractogram of the crystal image (figure 4(c)). The defocus value is roughly determined to be -39 nm by matching the intensity profile of the Thon diffractogram with several CTF curves.

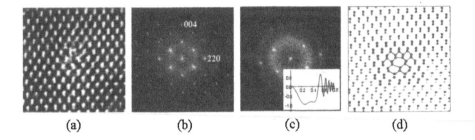

Figure 4. (a) [110] image of $Si_{0.76}Ge_{0.24}$/Si taken with a JEOL JEM 2010F microscope , (b) digital diffractogram from a large area including (a), (c) diffractogram from both the amorphous and crystalline area nearby the dislocation with the matched CTF curve as inset on the bottom right and (d) deconvoluted image corresponding to (a).

Seven trial $F'_{trial}(H)$ sets were obtained from the corresponding trial defocus values from -40 nm to -37.5 nm with defocus steps of 0.5 nm, and the corresponding corrected $F'_{trial}(H)$ sets were calculated subsequently. The inverse FT of the corrected $F'_{trial}(H)$ sets yielded seven potential maps. The best map (figure 4(d)), in which all atoms are resolved most clearly, was selected from among these seven maps. It can be seen that the atoms that form the core structure of a Lomer dislocation are resolved individually, with the bonding situation shown by linking the adjacent atoms. Each black dot represents the position of an atomic column projected in the [110] direction. A five-membered ring and a seven-membered ring form the symmetric undissociated core without a dangling bond. This core structure is in agreement with the Hornstra model [23]. It confirms that a Lomer dislocation can be formed with the Hornstra structure in the region of epilayers close to the interface boundary.

(2) 60° *dislocation complex* [15]

Figure 5(a) shows another [110] image of $Si_{0.76}Ge_{0.24}$/Si and figure 5(b) is the deconvoluted image corresponding to 5(a). The detailed geometry of the dislocation complex can be seen clearly by linking the adjacent atoms to show the bonding situation. It is seen that the dislocation complex is composed of a perfect 60° dislocation (left) and an extended 60° dislocation (right). The seven-membered rings and five-membered rings are observable in the core regions of both dislocations. The extended 60° dislocation dissociates into two partial dislocations of 90° and 30° with a short stacking fault between them.

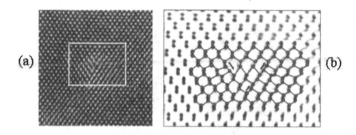

Figure 5. (a) [110] image of $Si_{0.76}Ge_{0.24}/Si$, (b) the deconvoluted image corresponding to the framed rectangular area in (a). Arrows in (b) denote the Burgers vectors.

Structures of dislocation core, micro-twin and interfacial boundary in 3C-SiC/Si [16]

(1) Recognition of Si and C atoms

Figure 6(a) shows a perfect region of a [110] image of cubic ($3C$)-SiC/Si. The crystal thickness increases from the left to the right. The corresponding deconvoluted image is given in figure 6(b), in which each pair of atomic columns of Si and C is revealed as a black dumbbell. In order to discern the contrast differences between the two ends of the dumbbells, the gray levels were measured for three regions I, II and III and the contrast profiles are shown in figure 6(c). Gray values from 0 to 255 were assigned in correspondence to the contrast change from white to black. The gray level for the left ends of the dumbbells is lower than, almost equal to, and slightly higher than that for the corresponding right ends, for regions I, II and III, respectively. Because region I is near the amorphous area and hence the thinnest one among these three regions, it is reasonable to approximate region I as a weak phase object, while region II and III should be treated as pseudo weak phase objects. The contrast of atomic columns of Si and C shown in figure 6(c) is in agreement with the pseudo WPOA [8]. The fact that with the increase of crystal thickness the peak height for light atoms goes up quicker than that for heavy atoms corresponds to the case below the critical crystal thickness. Therefore, the left ends of the dumbbells are recognized as atomic columns of C while the right ends are Si.

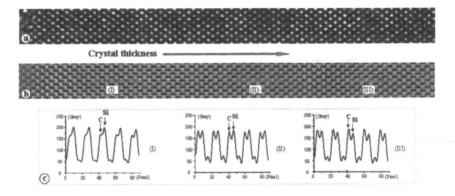

Figure 6. (a) Experimental image of 3C-SiC, (b) deconvoluted image corresponding to (a) and (c) profiles of gray levels measured for three regions I, II and III in (b).

(2) 30 °partial dislocation core structure

The micrograph shown in figure 7(a) is selected from the same [110] image as figure 6(a). A 30° partial dislocation (the arrow points to the terminated plane and the terminal is marked by a circle) associated with a stacking fault (SF) can be seen. Horizontal streaks caused by the SF are seen clearly on both sides of all reflections included in the digital diffractogram (see the inset on the top left of figure 7(a)). The deconvoluted image for figure 7(a) with Δf = -47 nm is shown in figure 7(b). The strong streaks seen in the digital diffractogram inserted in figure 7(b) indicates that most of the structural information contributed by the SF is retained. This is because the planar defect information is mostly retained by utilizing an elliptical window [24] for the Fourier filtering. In figure 7(b), all atoms appear as black and the dumbbells are recognizable (although not clearly). In figure 7(c) the structure model is superimposed on the deconvoluted image. Due to the limited resolution of the microscope, the two ends in the dumbbells are not resolved.

For a 30° partial dislocation, the inserted plane may terminate either in a Si atom or a C atom [25]. To identify the terminal atom marked by the small arrows in figures 7(b) and (c), the above-mentioned argument of image contrast variation with crystal thickness was employed. It can be justified from the matrix image for both figures 6 and 7 that the contrast profile in the segment III shown in figure 6(c) can be used for figure 7(c) to make out that the gray level for Si is lower than that for C. Thus the smaller ends in the dumbbells correspond to atomic columns of Si, while the bigger ends correspond to the atomic columns of C. Therefore, all Si and C atoms including those marked by the small arrows can be identified in the deconvoluted image shown in figure 7(c) so that it is clear that the partial dislocation terminates in the C atomic column. The atomic configuration of the 30° partial dislocation core and its associated SF was then derived and the corresponding model is superimposed on the deconvoluted image (see figure 7(c)). The stacking sequence is ···γaαβcγaα'c'γaαβcγa···, where αβγ (α') represents atomic layers of Si and abc(c') representing atomic layers of C.

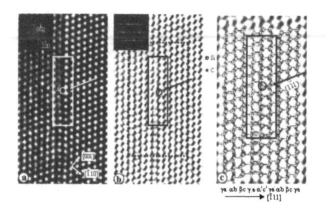

ya αb βc γ a α'c' ya αb βc γa

→ [Ī11]

Figure 7. (a) [110] image of 3C-SiC, (b) deconvoluted image corresponding to (a) and (c) structure model of 30° partial dislocation superimposed on the deconvoluted image.

(3) Structures of micro-twin and twin boundaries

In 3C-SiC, the {111} twins could be either a 180° rotation twin or a mirror reflection twin [26] with the corresponding stacking sequence ···αβbγcβbαa··· or ···αβbγγbβaα···, respectively. The micrograph given in figure 8(a) is from the same matrix image as figures 6(a) and 7(a). It contains a segment of {111} micro-twin (marked by two arrows). The corresponding diffractogram is inserted on the top right. Figure 8(b) shows the deconvoluted image corresponding to figure 8(a) with Δf = -46.5 nm. Again atoms appear black and the two atoms in the dumbbells are not resolved clearly. Since this region is near the thin amorphous area, the thickness should be comparable to that shown in the left part of figure 6(a). Hence the contrast profile of segment I of figure 6(c) can be used for figure 8(b). Thus it is reasonable to treat the bigger ends of the dumbbells as atomic columns of Si. The structure model of the micro-twin segment together with its boundaries constructed on the basis of the above argument is superimposed on the deconvoluted image (see figure 8(b)). It is seen that the stacking sequence is ···αβbγcαaγcβbαaγcββbγcααβb···, and the micro-twin segments is sandwiched between two 180° rotation twins. Figure 8(c) shows the image simulated according to the model given in figure 8(b) with Δf = -46.5 nm and a crystal thickness of 3.08 nm. The simulated image agrees well with the experimental image.

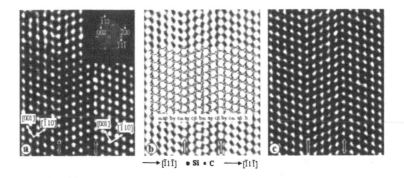

$\longrightarrow [\bar{1}1\bar{1}]$ • Si • C $\longrightarrow [\bar{1}1\bar{1}]$

Figure 8. (a) [110] image of 3C-SiC, (b) deconvoluted image corresponding to (a) with the micro-twin structure model superimposed and (c) image simulated based on the model given in (b).

CONCLUSIONS

This paper illustrates that image deconvolution is effective in enhancing HREM imaging for the determination of crystal defects and incommensurately modulated structures. The essence of the method is to transform a single image that is not intuitively representing the structure into a structure image and to improve the image resolution up to the information limit of the microscope. In some cases, it is important to recognize different atoms in the crystals on the basis of the pseudo WPOA. The deconvolution is applicable to images taken with microscopes that have different electron sources. In the case of 200 kV FEG microscopes, the image resolution has been improved from 0.2 nm to 0.14 nm. As for 200 kV microscopes fitted with LaB$_6$ filaments, though the information limit is not high, the image quality may be improve significantly, and much more important structural details are revealed in the images after the deconvolution process.

ACKNOWLEDGEMENTS

The research project was supported by the National Natural Science Foundation of China (Grants No. 10874207 and No. 50672124).

REFERENCES

1. J. M. Cowley and S. Iijima, *Z. Naturforsch.* **A27**, 445 (1972).
2. N. Uyeda, T. Kobayashi, Y. Suito, and M. Harada, *J. Appl. Phys.* **43**, 5189 (1972).
3. O. Scherzer, *J. Appl. Phys.* **20**, 20 (1949).
4. P. Schiske, in *Proceedings 4th European Conference on Electron Microscopy* (Rome, 1968) pp.

145-146.

5. D. van Dyck, H. Lichte, and K. D. Van der Mast, *Ultramicroscopy* **64**, 1 (1996).

6. F. H. Li, *J. Microscopy* **190**, 249 (1998).

7. F. H. Li, *Zeitschrift fur Kristallographie* **218**, 279 (2003).

8. F. H. Li and D. Tang, *Acta Cryst.* **A41**, 376 (1985).

9. F. H. Li, D. Wang and W. Z. He, and H. Jiang, *J. Electron Microsc.* **49**, 17 (2000).

10. J. J. Hu and F. H. Li, *Ultramicroscopy* **35**, 339 (1991).

11. D. X. Huang, W. Z. He, and F. H. Li, *Ultramicroscopy* **62**, 141 (1996).

12. H. B. Wang, Y. M. Wang, and F. H. Li, *Ultramicroscopy* **99**, 165 (2004).

13. F. Thon, *"Phase contrast electron microscopy,"* *Electron Microscopy in Material Science*, ed. U. Valdré (Academic press, New York and London, 1971) pp. 570-625.

14. D. Wang, H. Chen, F. H. Li, K. Kawasaki, and T. Oikawa, *Ultramicroscopy* **93**, 139 (2002).

15. D. Wang, J. Zou, H. Chen, F. H. Li, K. Kawasaki, and T. Oikawa, *Ultramicroscopy* **98**, 259 (2004).

16. C. Y. Tang, R. Wang, F. H. Li, X. H. Zheng, and J. W. Liang, *Phys. Rev.* **B75**, 184103 (2007).

17. C. Wen, W. Wan, F. H. Li, Z. H. Li, J. M. Zhou, and H. Chen, *Proceedings of the 9th Asian-Pacific Microscopy Conference* (Jeju, Korea 2008), 531-532.

18. W. Wan, C. Y. Tang, Y. M. Wang and F.H. Li, Acta Physica Sinica. **54**(9), 4273 (2005)

19. Y. M. Wang, W. Wan, R. Wang, F. H. Li and G. C. Che, *Philosophical Magazine Letters* **88**(3), 481(2008).

20. B. H. Ge, Y. M. Wang, X. M. Wang, and F. H. Li, *Philosophical Magazine Letters* **88**(3), 213 (2008).

21. F. H. Li, X. M. Li and B. Ge, *Proceedings of the 9th Asian-Pacific Microscopy Conference* (Jeju, Korea 2008), 137-138.

22. A. Bourret, J. Desseaux, and A. Renault, *Phil. Mag.* **A45**, 1 (1982).

23. A. J. McGibbon, S. J. Pennycook, and J. E. Angelo, *Science* **269**, 519 (1995).

24. C. Y. Tang and F H Li, *J. Electron Microscopy* **54**, 445 (2005).

25. A. T. Blumenau, C. J. Fall, R. Jones, S. Öberg, T. Frauenheim, and P. R. Briddon, *Phys. Rev.* **B68**, 174108 (2003).

26. D. B. Holt, *J. Phys. Chem. Solids* **25**, 1385 (1964).

Mater. Res. Soc. Symp. Proc. Vol. 1184 © 2009 Materials Research Society   1184-GG03-02

Quantitative Structure Analysis of Nanosized Materials by Transmission Electron Microscopy

Wolfgang Neumann, Holm Kirmse, Ines Häusler, Changlin Zheng, Anna Mogilatenko
Humboldt University of Berlin, Institute of Physics, Chair of Crystallography, Newtonstrasse 15, 12489 Berlin, Germany

ABSTRACT

The quantitative analysis of nanostructured materials increasingly requires the combined use of a variety of complementary electron microscopical techniques as well as new interpretation techniques as feature sizes decrease. The following studies of quantitative TEM analysis will illustrate this statement. The three-dimensional shape of (Si,Ge) semiconductor islands grown by liquid phase epitaxy (LPE) on Si substrates was determined by electron holography. The chemical composition of the islands was determined by quantitative high resolution electron microscopy (qHRTEM) and energy dispersive X-ray spectroscopy (EDXS). ZnTe nanowires and the Au-based catalyst droplet grown on GaAs via a vapour-liquid-solid (VLS) process were characterized by HRTEM, EDXS and electron energy loss spectroscopy (EELS). The possibilities of composition analysis of ternary semiconductors by combined application of conventional TEM (dark-field imaging) and HRTEM were demonstrated for the determination of the antimony content in Ga(Sb,As) quantum dots (QDs) grown by metal organic vapour deposition on GaAs substrates. Additionally, studies of chemically sensitive imaging of this QD-system by means of scanning transmission electron microscopy (STEM) will be discussed. The magnetic domain structure of soft magnetic FeCo based nanocrystalline alloys was investigated by Lorentz microscopy and off-axis electron holography. A strong correlation between the microstructure of the alloys and the structure of the magnetic domains was found.

INTRODUCTION

Nanostructured materials from almost all classes of materials are of great interest because the reduced dimensionality may drastically change the physical properties. In general these properties are a function of size, shape, arrangement, structure and chemical composition of the nanosized materials. Transmission electron microscopy (TEM) allows a detailed insight into the material characteristics. In order to correlate microstructure, microchemistry and materials properties the various TEM techniques for imaging, diffraction and spectroscopy have to be combined.

The classical diffraction contrast method of conventional TEM is applied to analyse the size, shape and arrangement of nanosized structures, where a quantitative analysis often requires image simulations of diffraction contrast for theoretical structure models. An alternative and powerful method is the three-dimensional reconstruction of the shape of nanostructures from two-dimensional phase mapping by means of electron holography. Furthermore, electron holography and Lorentz microscopy are useful for the evaluation of structure/magnetic property relationships. Quantitative high-resolution transmission electron microscopy (qHRTEM) provides information on structure and chemical composition at an atomic scale of magnitude.

Analytical TEM (energy-dispersive X-ray spectroscopy (EDXS), electron energy-loss spectroscopy (EELS), energy-filtered TEM (EFTEM)) and STEM (scanning TEM) Z-contrast imaging can be applied for the direct determination of chemical composition.

It should be noted that the various quantification methods have to be modified very often for the analysis of nanostructures with respect to their individual geometry. In this paper we present selected results of quantitative TEM analysis of semiconductor islands, nanowires, quantum dots and soft magnetic materials.

QUANTITATIVE ANALYSIS

(Si,Ge) islands on Si

(Si,Ge) is an appropriate model system for investigating the self-organisation of nanostructures in semiconductor heteroepitaxy. The $Si_{1-x}Ge_x$ islands were grown on (001) Si by liquid phase epitaxy (LPE) using different growth procedures. Size, shape and arrangement of the (Si,Ge) islands were investigated by conventional TEM of plan-view and cross-section specimens in the diffraction contrast imaging mode. In order to get information on the 3D-shape of an arrangement of islands simultaneously off-axis electron holography was applied [1]. An electron hologram, the reconstructed amplitude and phase image of $Si_{0.6}Ge_{0.4}$ islands are shown in Fig. 1. The phase shift of the electron wave modulated by the inner potential of the material is given by:

$$\Delta\varphi = C_E V_0 t,$$

where C_E is the interaction constant depending on the accelerating voltage, V_0 is the mean inner potential and t is the specimen thickness. If the mean inner potential is known, the 3D-shape of the islands can be reconstructed from the 2D-phase mapping.

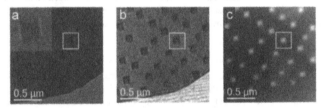

Fig. 1: (a) Electron hologram of $Si_{0.6}Ge_{0.4}$ islands on (001) Si substrate (insert: enlarged island). (b) Reconstructed amplitude image and (c) unwrapped phase image.

Fig. 2: Reconstructed 3D-image of $Si_{0.6}Ge_{0.4}$ islands on Si (001) substrate.

The mean inner potential of (Si,Ge) alloy was derived from a linear fitting of the data of Si (12.57 V) and of Ge (14.67 V), which were calculated by P Kruse et al [2] using density functional theory. The reconstructed image of an array of islands is shown in Fig. 2.

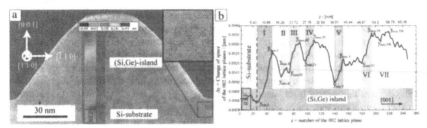

Fig. 3: Quantification of composition and strain of (Si,Ge) islands: a) c_S-corrector HRTEM image of a (Si,Ge) island with the coloured map of the 2D-displacement field; b) derivative of the displacement field in [001] direction of the colour-coded area of Fig. 3a.

In order to quantify strain in the $Si_{0.6}Ge_{0.4}$ islands HRTEM analysis of a cross-sectional specimen has been performed. A representative HRTEM image of a (Si,Ge) island viewed in the [110] zone axis is shown in Fig. 3a. In the field of view the specimen thickness is approximately 15 nm as revealed by tilting experiments in the TEM. Strain state of the (Si,Ge) islands was analyzed by the peak finding method DALI [3] . The colour-coded map superimposed with the HRTEM image in Fig. 3a shows the two-dimensional displacement field u_z of the atom columns measured with respect to a reference area (REF), which was chosen in the unstrained Si substrate underneath the (Si,Ge) island. The derivative Δy of the displacement field u_z with respect to the position z along the [001] direction corresponds to a change of the 002 lattice parameter along the [001] growth direction (Fig. 3b).

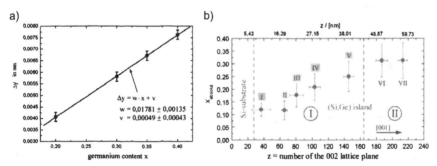

Fig. 4: Quantification of Ge composition of (Si,Ge) islands: a) gradient of the displacement as a function of the Ge content at the interface island/substrate; b) Ge concentration as a function of the island height. The Roman numerals correspond to the numerals of Fig. 3b.

To yield quantitative information from this curve it is necessary to know a correlation between the 002 lattice plane change and the Ge content. This correlation was calculated based on finite element method simulations of the (Si,Ge) islands with different Ge composition (Fig. 4a) [4]. Figure 4b shows the calculated Ge profile along the island height. These results show that the (Si,Ge) island consists of two regions showing a different Ge distribution. In the lower part of the island the Ge concentration continuously increases in the [001] direction up to about 25 at.%, whereas in the upper part of the island it remains constant at 30 at.%.

In addition to the qHRTEM analysis the chemical composition of the islands was directly determined by means of EDXS. Fig. 5 shows the result of an EDXS line-profile analysis obtained from a $Si_{0.6}Ge_{0.4}$ island. As visible the thin film approach which is routinely used in EDXS for composition quantification of thin TEM specimens can not give a reliable quantitative information in this case. The measured qualitative data are mainly affected by the specimen thickness decreasing linearly from island base (ca. 120 nm) to island top (ca. 50 nm). The aspect of thickness variation was taken into account and a modified ZAF correction method was developed and applied providing the quantitative results (solid line in Fig. 5b). In consistence with the above shown qHRTEM studies the EDXS analysis clearly exhibits an increase of Ge until 25 at.% is reached. The Ge content of the upper part of the islands remains constant. The results are also in good agreement with EFTEM and X-ray measurements.

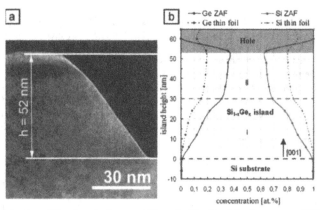

Fig. 5 Determination of the chemical composition of a $Si_{1-x}Ge_x$ island grown on (001) oriented Si substrate: (a) Cross-sectional dark-field TEM image; (b) Ge and Si concentration determined by EDXS analysis.

Quantitative TEM analysis of ZnTe nanowires

Semiconductor nanowires (NWRs) are objects of a length of several micrometers and of a diameter of only about 50 nm. The ZnTe NWRs under investigation are grown on GaAs substrate at 450 °C via a vapour-liquid-solid (VLS) process realized in a molecular beam epitaxial system [5]. Nanosized droplets of an Au-based catalyst promote the one-dimensional growth. For TEM analysis the NWRs were harvested from the substrate and transferred to a commercial carbon film support grid.

In Fig. 6 an overview image is given of a single ZnTe NWR. The Au-based catalyst droplet is clearly visible at the NWR tip. The ⟨111⟩ growth direction is marked by an arrow. The side wall of the NWR appears to be flat near to the tip. Here, only one-dimensional growth is realized and the cross section amounting to about 50 nm in diameter is determined by the size as well as the shape of the catalyst droplet. In a larger distance from the tip the NWR is tapered and facets point out of the side wall. Both features evidence a lateral growth starting where the attraction of the catalyst no longer forces the entire number of atoms to move to the droplet.

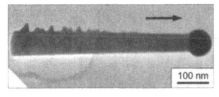

Fig. 6: Overview image of a harvested ZnTe NWR transferred to a carbon film support. The arrow marks the growth direction ⟨111⟩.

In order to gain a more thorough understanding of the VLS growth process the catalyst droplet was examined [6]. An amorphous layer covering the catalyst droplet was found. The thickness of the layer amounts to about 3 nm. The origin of this layer can be understood by quantitative analysis of the chemical composition by means of EDXS (cf. Fig. 7). For the analysis the thin film approximation was applied.

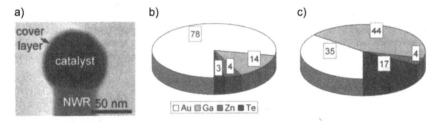

Fig. 7: Analysis of the Au-based catalyst droplet: a) TEM image of the tip of a NWR; b) and c) quantified chemical composition of the droplet (b) and of its cover layer (c) as revealed by EDXS and successive composition evaluation.

The composition of the catalyst droplet is given in Fig. 7b. As expected the droplet contains a majority of Au (78 at.%). The high amount of Ga (14 at.%) is due to the formation process of the catalyst droplets comprising deposition of a thin Au layer on the GaAs substrate and consecutive reorganization in droplets by an annealing step. During annealing Ga is dissolved from the substrate. The amount of remnant Zn and Te is 3 and 4 at.%, respectively.

Contrary to that, the cover layer contains 17 at.% of Te (Fig. 7c). On the average the content of Te in the droplet is higher compared to Zn proving the limitation of the NWR growth by the provision of Zn. The striking difference of the composition of the droplet and cover layer is due to strong segregation of Ga and Te while lowering the substrate temperature down to the melting point of the Au-based catalyst.

The crystalline structure of the NWRs is visualized in Fig. 8. The HRTEM image was taken from the sidewall of the NWR. The zone axis is [01$\overline{1}$]. Numerous stacking faults (SF) separated by a distance of only a few monolayers are seen. The SFs were identified to be extrinsic, where in contrast to the face-centred cubic stacking (…AαBβCγAαBβCγ…) a stacking sequence of …AαBβCγBβAαBβCγ… is found. Moreover, rotational twins (T) with [111] rotational axis were recognized.

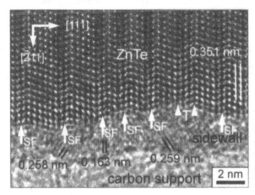

Fig. 8: HRTEM image of the NWR and its sidewall. Structural defects as stacking faults (SF) and rotational twins (T) are found in the NWR. The sidewall contains nanocrystals. The quantified lattice plane distances are given.

Similar to the catalyst droplet the sidewall of the NWR is covered by a thin layer. From the lattice plane features seen in Fig. 8 it is evident that there are nanocrystals within this layer which are not observed in the cover layer of the catalyst. Quantifying the interplanar distances a good agreement was found with lattice planes distances tabulated for hexagonal ZnO. Probably the ZnO shell was formed when the NWRs get in contact with atmospheric oxygen after taking the NWRs out of the growth chamber.

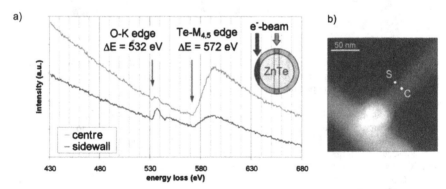

Fig. 9: Electron energy loss spectroscopy of a harvested ZnTe NWR: a) EEL spectra of the centre (C) and the sidewall (S). Up right a scheme is inserted for illustrating the contribution of the cover layer to the EEL signal for different positions of the electron beam; b) HAADF image of the NWR with markers of the points of EELS analysis.

In order to verify the presence of oxygen within the cover layer EELS was performed (Fig. 9). For comparison spectra were acquired at the surface (S) of the NWR and at the centre

(C) (see Fig. 9a). In the high angle annular dark-field (HAADF) image of Fig. 9b the positions of the EELS measurements are marked. Inspecting the O-K ionisation edge, the height of the peak is increased at the surface of the NWR compared to the centre position. In contrast to that, the Te peak intensity decreases by a factor of 3. In addition, the Zn-$L_{2,3}$ edge was analysed (not shown here). For both positions the height of the Zn peak remains constant.

This behaviour can be explained by a core/shell geometry of the NWR as shown in the inset of Fig. 9a. Here, the contribution of an atomic species which is only present in the sidewall is low for the centre position and reaches its maximum at the sidewall. From the EELS experiments the following elemental distribution has to be concluded. Te is present only in the NWR and not in the sidewall, Zn is found in both regions and O is only within the sidewall. In combination with the lattice plane distances determined for the nanocrystals from HRTEM images the formation of ZnO in the sidewall has to be concluded.

This example shows that information on the geometry of nanostructures is crucial for quantifying the chemical composition basing on analytical methods.

Ga(Sb,As) quantum dots embedded in GaAs

Self-organised Ga(Sb,As)/GaAs quantum dots (QDs) formed via the Stranski-Krastanov mode [7] were grown by metal organic chemical vapour deposition at 470°C. Depending on the growth conditions QD sizes ranging from about 5 nm to 40 nm were obtained. Methods of qHRTEM were used to determine the local strain field and chemical composition of In and Sb on atomic scale.

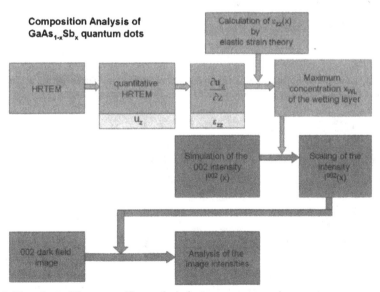

Fig. 10: Flow chart of the composition analysis for ternary compounds.

We have developed a quantitative method for the determination of chemical composition of ternary semiconductors comprises the combined use of dark-field imaging and qHRTEM [8]. The flow chart of this method is given in Fig. 10. The boundary conditions for the applicability of the method are the following: First, the diffraction patterns of the material system must contain chemically sensitive reflections. Second, the matrix material adjacent to the ternary layer has to contain only atomic species which are also components of the layer.

An example of the determination of the Sb content of a Ga(Sb,As) QD embedded in GaAs is illustrated in Figs. 11 and 12. The chemically sensitive 002 dark-field image of a Ga(Sb,As) QD layer embedded in GaAs is shown in Fig. 11a. For this image the specimen was oriented 6° off the [100] zone axis towards [110] while tilted towards [001] less than 1° until the optimum excitation of the 002 beam was achieved.

Fig. 12 shows a HRTEM image for [100] zone axis orientation which was analysed by peak finding method (DALI program package, [3]) to visualise the strain distribution across the Ga(Sb,As) wetting layer. The superimposed colour-coded map exhibits the measured displacement field u_z of the atomic columns in growth direction [001] compared to an unstrained reference area. Basing on this, the antimony concentration of the QD layer was calculated according to the procedure shown in Fig. 10 (see Fig. 11b).

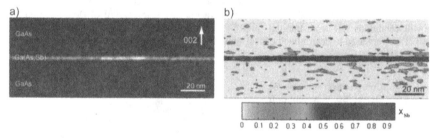

Fig. 11: Evaluation of Sb content of a Ga(Sb,As) QD embedded in a GaAs matrix: a) chemically sensitive 002 dark-field image; b) colour-coded map of Sb concentration x_{Sb}.

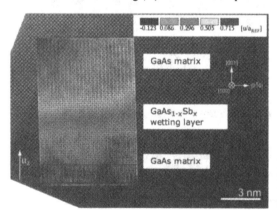

Fig. 12: C_s-corrected HRTEM image of Ga(Sb,As)/GaAs system taken along the [100] direction (REF – unstrained reference area).

For a direct visualisation of the differently composed QD layers and even of the individual atomic columns the STEM HAADF image of Fig. 13 was recorded in a probe-c_S-corrected TEM/STEM. With corrected c_S a lateral resolution below 0.1 nm is realized.

The regularly arranged bright dots seen in Fig. 13a correspond to the positions of individual atomic columns viewed along the [100] direction of the sphalerite crystal structure. Within the field of view lattice planes are seen exhibiting a higher brightness compared to the GaAs regions. The brightness difference corresponds to an increase of the mean atomic number of the transmitted material. Fig. 13b gives three line profiles for the quantification of the QD layer thickness and the identification of the different atomic columns. The blue curve corresponds to the experimental HAADF intensity measured along the arrow given in Fig. 13a. Within the framed area the intensity was averaged perpendicular to the arrow. For the blue curve a background signal can be defined (cf. dashed line in Fig. 13b). This signal is related to the specimen thickness. The linear decrease parallel to the path of analysis hints to a wedge-shaped specimen after preparation for TEM. The background intensity at the layer positions is increased due to an additional contribution (see dotted-dashed line). Here, two origins have to be considered. First, the crystal lattice is strained due to the lattice mismatch between (In,Ga)As and GaAs. Second, the efficiency of scattering at positions in between the atomic columns might be increased when heavier atoms are incorporated. The asymmetric shape of the triangle marked at the (In,Ga)As layer has to be attributed to the segregation of In atoms.

a)   b)

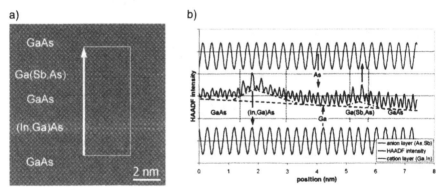

Fig. 13: Chemically sensitive imaging of an (In,Ga)As and a Ga(Sb,As) QD layer embedded in GaAs: a) HAADF image acquired with corrected c_S for the electron probe; b) line profile of HAADF intensity (blue line) as detected within the area framed in a). The sequences of the cations (Ga and In, cf. magenta line) and of the anions (As and Sb, cf. green line) are given. The dashed line marks the background intensity due to a thickness wedge of the TEM specimen. The dotted-dashed line symbolizes the intensity contribution due to the modified composition.

In addition, the curve shows numerous oscillations correlated with the lattice planes of cations and anions, respectively. For GaAs the signal originating from the As columns is brighter than that of the Ga columns. Correspondingly the magenta (cations) and the green curves (anions) are given.

The (In,Ga)As layer is detectable from the increased HAADF intensity of the cation positions. For about 5 cation-anion double layers the intensity is altered compared to GaAs. This layer thickness corresponds to 1.5 nm. For Ga(Sb,As) the anion positions are expected to reveal a higher brightness. Indeed, as visible in Fig. 13b the higher peaks in the blue curve at the Ga(Sb,As) layer correlate with the anion positions. The Ga(Sb,As) layer is only 2 double layers (0.6 nm) thick.

Softmagnetic FeCo-based alloy

Nanocrystalline soft magnetic alloys are promising materials for power conversion applications due to the low coercivity and high permeability. In order to have a thorough understanding of the soft magnetic properties and for tailoring the magnetic properties the structure/property relationships in nanocrystalline soft magnetic alloys have to be investigated. Additionally to the structure analysis by means of CTEM and HRTEM, electron holography was used to image the magnetic structures with high resolution and high sensitivity.

FeCo-based nanocrystalline alloys ($(Fe_{0.5}Co_{0.5})_{80}Nb_4B_{13}Ge_2Cu_1$) were grown by annealing an amorphous ribbon (fabricated by melting spinning) at 500°C for one hour. The annealed alloy was prepared for TEM analysis by mechanical dimpling and consecutive Ar-ion milling. In order to prevent the objective lens field to destroy the magnetic domain structures, both Lorentz microscopy and electron holography investigations were performed in Lorentz mode.

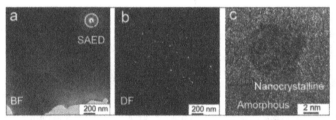

Fig. 14: Microstructure of $(Fe_{0.5}Co_{0.5})_{80}Nb_4B_{13}Ge_2Cu_1$ nanocrystalline soft magnetic alloy: (a) bright-field TEM image (inserted: diffraction pattern); (b) dark-field TEM image; (c) HRTEM image.

Microstructure investigations indicate that the formed alloy is a two-phase structure consisting of small α-FeCo crystalline particles (bcc structure) embedded in the residual amorphous matrix (Fig. 14). The FeCo particles are randomly oriented and the average grain size is ~12 nm.

Further magnetic domain structure analysis by means of Lorentz microscopy and electron holography indicates that the alloy contains large magnetic domains up to several micrometers in size with few pinning sites (Fig. 15). In addition, a dynamical magnetization experiment was performed by tilting the sample in a weakly excited objective lens field, as shown in Fig. 16. The reconstructed phase images indicate that the alloys can be easily magnetized and demagnetized in a weak applied field hinting to good soft magnetic properties.

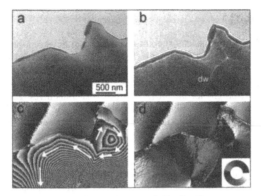

Fig. 15: Magnetic domain structure of $(Fe_{0.5}Co_{0.5})_{80}Nb_4B_{13}Ge_2Cu_1$ nanocrystalline soft magnetic alloy: (a) in-focus and (b) defocused Lorentz microscopy images; (c) reconstructed electron holography phase image; (d) colour-coded mapped magnetic domain structures. The magnetic flux directions are represented by the inserted colour wheel.

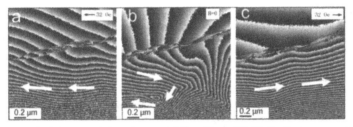

Fig. 16: Series of reconstructed phase images of $(Fe_{0.5}Co_{0.5})_{80}Nb_4B_{13}Ge_2Cu_1$ nanocrystalline soft magnetic alloy while tilting the sample in a weakly excited objective lens field.

Such structure/property relationships can be explained by the Herzer random anisotropy model [9]. The randomly oriented FeCo particles are ferromagnetically coupled by exchange interaction through the interfacial amorphous matrix. Then, the effective magnetocrystalline anisotropy is significantly reduced by the average effect leading to good soft magnetic properties.

CONCLUSIONS

We have demonstrated how the various methods of TEM can be applied to characterize nanostructured materials. The continued development and application of both the analytical techniques and the interpretation methods for the analysis of nanostructures as a function of dimensionality is essential.

ACKNOWLEDGMENTS

We acknowledge the financial support of the German Science Foundation (DFG). Many thanks is given to H. Wawra, Dr Th. Boeck (Leibniz Institute of Crystal Growth Berlin), Prof T. Wojtowicz (PAS Institute of Physics, Warsaw), Dr L. Müller-Kirsch, Prof D. Bimberg (Technical University Berlin), Prof M. E. McHenry, Prof D. E. Laughlin, Dr J. Long (Carnegie Mellon University, Pittsburgh) for the provision of the materials investigated. Thanks are given to the JEOL company for the possibility to investigate the Ga(Sb,As) QD layer system in a probe-c_S-corrected TEM/STEM.

REFERENCES

1   Ch. Zheng, H. Kirmse, I. Häusler, K. Scheerschmidt and W. Neumann, Proc. 14th Europ. Micr. Congr., 2008, Aachen (Germany), Vol. 1: Instrumentations and methods, ed.: M. Luysberg, K. Tillmann, Th. Weirich, 287-288.

2.   P. Kruse, M. Schowalter, D. Lamoen, A. Rosenauer and D. Gerthsen, Ultramicroscopy **106**, 105 (2006).

3.   A. Rosenauer, S. Kaiser, T. Reisinger, J. Zweck, W. Gebhardt and D. Gerthsen, Optik **101**, 1 (1996).

4.   R. Köhler, W. Neumann, M. Schmidbauer, M. Hanke, D. Grigoriev, P. Schäfer, H. Kirmse, I. Häusler and R. Schneider, "Structural characterisation of quantum dots by X-ray diffraction and TEM", Semiconductor Nanostructures, ed. D. Bimberg (Springer, 2008) pp. 97 – 121.

5.   E. Janik, J. Sadowski, P. Dłużewski, S. Kret, L.T. Baczewski, A. Petroutchik, E. Łusakowska, J. Wróbel, W. Zaleszczyk, G. Karczewski, T. Wojtowicz, A. Presz, Appl. Phys. Lett. **89**, 133114 (2006).

6.   H. Kirmse, W. Neumann, S. Kret, P. Dluzewski, E. Janik, G. Karzewski and T. Wojtowicz, phys. stat. sol. (c) **5**, 3780 (2008).

7.   I.N. Stranski, L. Krastanow; Sitzungsberichte d. Akad. d. Wissenschaften in Wien, Abt. IIb,146, 1937, S.797

8.   I. Häusler, H. Kirmse, W. Neumann, phys. stat. sol. (a) **205**, 2598 (2008).

9.   G. Herzer. Physica Scripta. **T49**, 307 (1993).

Mater. Res. Soc. Symp. Proc. Vol. 1184 © 2009 Materials Research Society   1184-GG02-01

Reflection High Energy Electron Diffraction (RHEED) Study of Nanostructures: From Diffraction Patterns to *Surface* Pole Figures

Fu Tang, Toh-Ming Lu, and Gwo-Ching Wang
Department of Physics, Applied Physics and Astronomy, Rensselaer Polytechnic Institute
110, 8th Street, Troy, NY 12180

ABSTRACT

In this report we present a brief overview of the growth of nanostructures by the oblique angle deposition where the nanostructures possess both out-of-plane and in-plane preferred orientations or a biaxial texture. The degree of preferred crystal orientations can be quantitatively determined from a method called "RHEED surface pole figure analysis" that we developed recently.

INTRODUCTION

Many thin film materials and nanostructures are not single crystals but polycrystallines with a preferred orientation of the crystalline grains. There are various categories of common textures: fiber texture with one-degree orientation and biaxial texture with two-degree orientation. Extreme cases of texture are random orientations and single crystals. Many physical properties of materials depend on grain orientations. Therefore it is critical to understand the science of the texture evolution process in order to control and design its texture.

It is well known that reflection high-energy electron diffraction (RHEED) has been an important tool to monitor the growth rate of epitaxial films on single crystal substrates through "intensity oscillations" and to determine the structure of the films through the diffraction patterns. The questions are: what information can we obtain from RHEED of nanostructures grown on non-single crystal substrates such as amorphous substrates? Can we use this common laboratory electron diffraction technique to study the initial to the final stages of nanostructure growth in terms of their structure and texture? By understanding the science of growth, can we develop a method to grow single crystal nanostructures on an amorphous substrate that would change the ways of growing single crystal films?

Very often, the initial growth of nanostructures on an amorphous substrate exhibits a completely random crystal orientation which gives a ring structure in the RHEED pattern. From the ring radius and the full-width-at-half-maximum of the ring the lattice constant and the average size of nanocrystals (as small as ~2 nm) can be determined, respectively [1]. If these nanocrystals evolve into a preferred orientation the rings break and arcs are formed. An example of *in situ* RHEED pattern measurements of the growth of Cu nanocrystals on an amorphous substrate has been demonstrated in our lab. [1, 2]

A particularly interesting case is the growth of the nanostructures by the oblique angle deposition where the nanostructures possess both out-of-plane and in-plane preferred orientations or a biaxial texture. The degree of preferred crystal orientations can be quantitatively determined from a method called "RHEED surface pole figure analysis" that we developed recently [3, 4].

EXPERIMENT

Oblique angle deposition and shadowing effect

Figure 1(a) shows a schematic of the oblique angle vapor deposition technique. The oblique incident flux with respect to the substrate normal breaks the deposition symmetry as compared to that of the normal incidence where $\theta = 0°$. For a large incident angle θ (>85°), it is called glancing angle deposition, or GLAD [5]. The physical shadowing effect associated with the oblique angle flux creates isolated nanostructures because the tallest islands block the oblique incident flux and create a shadowing length as seen in Fig. 1(b). Within this length scale an empty space is created.

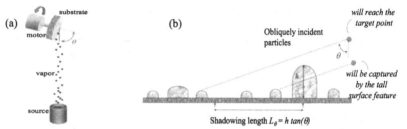

Figure 1. (a) A schematic of oblique angle deposition with substrate rotation. The source could be a sputtering target or thermal vapor. (b) This figure illustrates the effect of shadowing during oblique-angle deposition. Islands of different height are initially nucleated at the surface. Subsequently the incident flux of material that strikes the surface with an oblique angle α is preferentially deposited onto the top of surface features with larger height values.

RHEED surface pole figure setup and substrate rotation

Figure 2 shows our RHEED setup with substrate rotation. This is a homemade in situ growth and

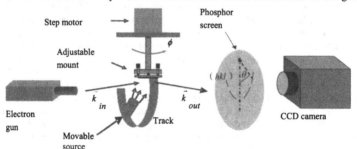

Figure 2. A schematic of the experimental setup for RHEED surface pole figure. A UHV step motor is used to rotate the substrate in-plane to obtain different slices of the pole figure. The angle ϕ is the azimuthal angle around the substrate normal. The angle θ is polar angle measured from the substrate normal. For *in situ* characterization an evaporation source was added in the UHV chamber, as shown in (a). The evaporation source can move along a semi-circular track allowing the deposition angles to be varied. [4]

diffraction measurement system. The incident electron energy is ~9 keV. The incident angle can be adjusted and is less than 2° with respect to the substrate surface plane. The CCD camera and its software were installed using commercial components [1]. This RHEED setup has a few advantages for *in situ* study: the measurement geometry prevents the glancing-angle incoming electron beam from interfering with the evaporation flux, and data acquisition has a high temporal resolution. The step motor that controls the rotation of substrate has a step size of 1.8°. A total of 200 patterns covering the 360° azimuthal angle were recorded for the pole figure construction. The exposure time for each image was 5 seconds. The total data collection time for 200 patterns was ~20 min [3]. For *in situ* characterization a thermal evaporation source was added in the UHV chamber, see Fig. 2.

Construction of RHEED surface pole figure

In RHEED we use a phosphor screen as two-dimensional detector to collect diffraction patterns. The use of a two-dimensional area detector to construct pole figures has been reported in X-ray diffraction (XRD) [6-9] and in transmission electron microscopy (TEM) operated in both transmission and reflection modes [10]. Compared with a point detector, the two dimensional detector can collect a large range of polar angles at each azimuthal angle. This would tremendously reduce the data collection time. Figure 3(a) shows a schematic to demonstrate the principle of using a two-dimensional detector to construct a pole figure. The grey sphere in Fig. 3(a) represents the reciprocal space of a family of crystal planes, for example, the (*hkl*) plane. The film is assumed to have a biaxial orientation. In the measurement, the Bragg diffraction for the (*hkl*) plane can only be satisfied when the amplitude of the difference between k_{in} and k_{out} is equal to the radius of the (*hkl*) reciprocal sphere. Then the angle between k_{in} and k_{out} is equal to the specific Bragg angle $2\theta_{(hkl)}$. For a particular k_{in}, these points are distributed along the bold dashed circle, or Debye ring. From the figure, we can see that this circle almost covers a full range of the polar angle θ, from 0° to 90°, at a fixed azimuthal angle ϕ.

Using a two dimensional detector, it is possible to record the whole (*hkl*) diffraction ring, shown in Fig. 3(a), to construct the (*hkl*) pole figure. Then the substrate just needs to be rotated azimuthally in order to record the diffraction intensities at various space orientations. In the

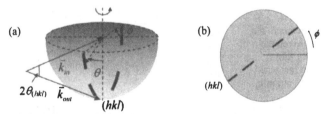

Figure 3. A schematic to demonstrate the principle of using two-dimensional detector to construct a pole figure. The $\vec{k}_{in}$ and $\vec{k}_{out}$ are the wave vectors of incident and scattered electron beams, respectively. The grey sphere in (a) represents the reciprocal space of a family of (*hkl*) crystal plane. The specific Bragg angle is $2\theta_{(hkl)}$. For a particular k_{in}, the points satisfying the (*hkl*) Bragg diffractions are distributed along a bolded circle consisting of arcs shown in (a). The figure (b) demonstrates the relationship between a diffraction pattern and pole figure. The bolded straight line in (b) is the projection of (*hkl*) diffraction arcs. [4]

RHEED measurement, the scattered electron beam is confined in a small spatial angle, so the whole diffraction ring can be easily captured. However, for XRD, X-ray can be scattered into much larger spatial angles so that a 2D detector may not be able to cover the whole range of a particular Bragg scattering. Figure 3(b) demonstrates the relationship between a diffraction pattern and pole figure. By definition a pole figure is a stereographic projection that represents the variation of pole density distribution of a specific family of planes [11]. If the film does not have a random orientation but has a texture such as biaxial, then the continuous ring will be broken into arcs. If one projects the (*hkl*) diffraction arcs in a semi-circle in Fig. 3(a) to the equator plane highlighted by the dashed circle then this would correspond to a slice of the corresponding (*hkl*) pole figure. This slice is represented as the bold straight dashed line in Fig. 3(b). The projection of arcs will have discrete poles or a variation of pole density distribution. As the substrate is rotated azimuthally, different slices can be obtained to compose the whole pole figure.

DISCUSSION

Mg nanoblades grown by oblique angle deposition

An example of the measurements of surface texture evolution during the growth of unusual Mg nanostructures on an amorphous substrate using *in-situ* RHEED surface pole figure technique has been presented in the past and we give a brief highlight here [12]. The substrate used for deposition was a Si wafer with a thin layer of native oxide on the surface. The vapor incident angle with respect to the substrate normal is ~75°. The distance between the evaporation source and the substrate holder was approximately 10 cm. The source was resistively heated to a desired temperature of ~600 K for evaporation. The base pressure of the vacuum chamber was ~4 × 10^{-9} Torr. The morphologies and structures of the final Mg nanoblades were imaged *ex-situ* by a field emission SEM. See Fig. 4 for the side view of nanoblades. The thickness of blades is in the tens nm and the width of blades is in a couple hundred nm. The thickness of the final Mg film obtained from the cross sectional SEM images is ~2.1 μm. The thickness refers to the vertical distance between the substrate and the Mg film surface. The growth rate was determined to be ~43 nm/min.

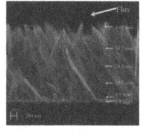

Figure 4. Side view SEM image of Mg nanoblades viewed from the direction perpendicular to the incident flux direction. The thicknesses corresponding to various deposition times are labeled by horizontal arrows. [12]

In situ monitoring of Mg nanoblades at a fixed azimuthal angle using RHEED patterns

During the Mg deposition the pressure rose to ~2.0 × 10^{-8} Torr. The Mg deposition was interrupted at various times from 0.5 to 34.7 min for *in-situ* RHEED pole figure measurements. Figure 5 shows RHEED patterns of Mg nanoblades when the electron beam is parallel to the wider width direction of nanoblades (sample position at $\phi = 0°(180°)$). Phi (ϕ) is the azimuthal angle around the substrate normal. Figures 5 (a), (b) and (c) are for deposition times 0.5 min (~22 nm thick), 13.7 min (~589 nm thick), and 34.7 min (~1.49 μm thick). The normal of ($10\bar{1}0$) planes and the [0001] axis are indicated by the white long dashed lines with arrows in (c). The short dotted line with an arrow in (b) and (c) represents the normal of $(10\bar{1}1)$ planes. For each film thickness we collected 200 RHEED patterns with $1.8°$ increment over $360°$ of azimuthal angle for surface pole figure analysis.

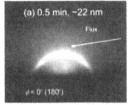 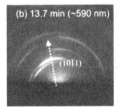 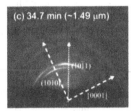

Figure 5. RHEED patterns of Mg nanoblades when the electron beam is parallel to the wider width direction of nanoblades (sample position at $\phi = 0°(180°)$). Deposition time: (a) 0.5 min (~22 nm thick), (b) 13.7 min (~589 nm thick), and (c) 34.7 min (~1.49 μm thick). The normal of $(10\bar{1}0)$ planes and the [0001] axis are indicated by the white long dashed lines with arrows in (c). The short dotted line with an arrow in (b) and (c) represents the normal of $(10\bar{1}1)$ planes. [12]

Biaxial texture evolution revealed from the surface pole figures

Figures 6 (a) to 6(d) show a series of normalized $(10\bar{1}1)$ RHEED pole figures at 0.5, 8.5 and 24.5 and 34.7 min deposition using the method described in the previous section. In the beginning of 0.5 min deposition the distribution of intensity in the pole figure intensity is nearly even, which indicates a random initial nucleation on the amorphous substrate. With more deposition at 8.5 min, an intense band is shown at the left side of the pole figure in Fig. 6(b). Clearly separated poles are revealed in the longer deposition time of 24.5 min (~1.05 μm) and 34.7 (~1.49 μm). The position of the poles in the figures moves towards the flux as the film grows. This indicates that the texture changes and the texture axis tilts more towards the flux. This change of texture axis is consistent with the evolution of polar intensity profiles, measured from RHEED images at $\phi = 0°(180°)$ shown in Fig. 5. We can see from Fig. 5(b) that at the deposition time of 13.7 min (~590 nm thick) the position of the peak with maximum intensity where the short dotted line passes through $(10\bar{1}1)$ arc is around $\theta = 20°$ on the side $\phi = 180°$. As the film grew thicker, the

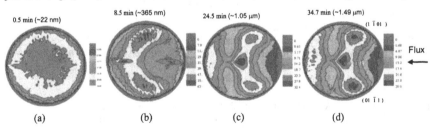

Figure 6. Normalized (1011) RHEED pole figures at deposition times (a) 0.5 min (~22 nm thick), (b) 8.5 min (~365 nm thick), (c) 24.5 min (~1.05 μm thick) and (d) 34.7 min (~1.49 μm thick). The positions of poles in the figures move towards the incident vapor flux as the film grows. [12]

peak position gradually moves to the center. At the deposition time of 34.7 min (~1.49 μm thick), the texture axis tilts to $\theta = 1°$ on the side $\phi = 0°$. From the trend, we can see that the texture axis tilting angle changes most dramatically at the early stage of growth. The x-ray pole figure of a 2.1 micron thick Mg nanoblade film generated from a laboratory x-ray source hardly shows any obvious pole. In contrast the RHEED surface pole of a 2.1 micron thick Mg nanoblade film is very obvious [4]. For a much thicker 10 micron Mg nanoblade film the x-ray pole becomes visible [4].

CONCLUSIONS

Since the electron used in RHEED has a sub-angstrom wavelength and a few nanometer penetration depth, we argue that it can probe the atomic structure and the crystal orientation of the growth front or near the film surface. It is in contrast with the conventional x-ray pole figure technique which gives average texture information of the entire film. In addition the scattering cross section of electron is orders of magnitude stronger than that of x-ray. Therefore the data collection speed is fast. The RHEED surface pole figure is suitable for *in situ* study and there is no need for a sample tilt mechanism like in the case of x-ray. From our Mg nanoblades we learn that texture evolution is a complex process and the texture of a film shown as rings or arcs in the RHEED patterns often changes during growth. Hopefully this information can shine some light on the basic mechanisms that control the film texture evolution. Research laboratories equipped with RHEED can in principle have the ability to perform the *RHEED surface pole figure* analysis by rotating the substrate to record multiple patterns diffracted from the surface.

ACKNOWLEDGMENTS

Work partially supported by the NSF 0506738.

REFERENCES

[1] J.T. Drotar, T.-M. Lu, and G.-C. Wang, J. Appl. Phys. **96**, 7071 (2004).
[2] F. Tang, C. Gaire, D.-X. Ye, T. Karabacak, T.-M. Lu, and G.-C. Wang, Phys. Rev. **B72** (3), 35430-1-8 (2005).
[3] F. Tang, G.-C. Wang, and T.-M. Lu, Appl. Phys. Lett. **89**, 241903 (2006).
[4] F. Tang, T. Parker, G.-C. Wang, and T.-M. Lu, J. Physics D: Appl. Phys. **40**, R427 (2007).
[5] K. Robbie and M.J. Brett, J. Vac. Sci. Technol. **A15**, 1460 (1997).
[6] K. Helming and U. Preckwinkel, Solid State Phenomena **105**, 71 (2005).
[7] S.L. Lee, D. Windover, M. Doxbeck, M. Nielsen, A. Kumar, and T.-M. Lu, Thin Solid Films **377**, 447 (2000).
[8] H.J. Bunge and H. Klein, Zeitschrift Fur Metallkunde **87**, 465 (1996).
[9] A. B. Rodriguez-Navarro, J. Appl. Cryst. **40**, 631 (2007).
[10] B. Schäfer and R.A. Schwarzer, Materials Science Forum **273–275**, 223 (1998).
[11] See, for example, *Elements of X-ray Diffraction*, B.D. Cullity, (Addison-Welsley, 1978), p. 297.
[12] F. Tang, G.-C. Wang, and T.-M. Lu, J. Appl. Phys. **102**, 014306 (2007).

Mater. Res. Soc. Symp. Proc. Vol. 1184 © 2009 Materials Research Society   1184-GG02-04

The Structural Phase Transition in Individual Vanadium Dioxide Nanoparticles

Felipe Rivera, Art Brown, Robert C. Davis, Richard R. Vanfleet
Brigham Young University – Department of Physics and Astronomy, N283 ESC,
Provo, UT 84602, U.S.A.

ABSTRACT

Vanadium dioxide (VO_2) single crystals undergo a structural first-order metal to insulator phase transition at approximately 68°C. This phase transition exhibits a resistivity change of up to 5 orders of magnitude in bulk specimens. We observe a 2-3 order of magnitude change in thin films of VO_2. Individual particles with sizes ranging from 50 to 250 nm were studied by means of Transmission Electron Microscopy (TEM). The structural transition for individual particles was observed as a function of temperature. Furthermore, the interface between grains was also studied. We present our current progress in understanding this phase transition for polycrystalline thin films of VO_2 from the view of individual particles.

INTRODUCTION

Vanadium is a transition metal whose oxides undergo a metal-to-insulator transition (MIT) at some characteristic temperature.[1,2] Among vanadium's oxides, vanadium dioxide (VO_2) has been extensively studied because its transition temperature occurs near room temperature (340K 68° C).[2–4] In single crystals, vanadium dioxide undergoes significant, abrupt, and reversible changes in several of its properties during its phase transition. These changes to its properties include: 1) A structural change from a low-temperature semi-conducting monoclinic phase to a high-temperature tetragonal metallic phase; 2) a resistivity change of several orders of magnitude; and 3) a sharp change in optical transmittance in the infrared region. These optical and electronic properties that vanadium dioxide exhibits due to its phase transition hint at the use of this material for optical[5–8] and electronic[9,10] applications, such as thermochromic coatings for windows, thermal sensors, or fast optical and electronic switches.

VO_2 shifts between a low-temperature, low-symmetry monoclinic structure to a high-temperature, higher-symmetry tetragonal structure.[11–14] Table I shows the standard structures for VO_2. Due to this structural change, it is possible to use scattering techniques to probe the transition of crystalline VO_2.[11–13,16] In this present study, Transmission Electron Microscopy (TEM) was employed to study this phase transition by taking advantage of the structural change in VO_2. Furthermore, Scanning Transmission Electron Microscopy (STEM) was used to probe the phase transition of individual grains as well as the interface between grains.

Even though the phase transition for VO_2 is cited to be 68°C (it was first induced by temperature [1]) there are several factors that will modify, tune, alter, or even induce the phase transition. Variations in stoichiometry, particle size, stress, misorientations between grains, morphological faults, dopants, and other "imperfections" have been used qualitatively to describe changes in the transition temperature, hysteresis, and sharpness of the transition.[8,16–18] However, despite the work that has been done on this material, the nature of the phase transition still is very much debated.[3,19–23] Therefore, in order to understand how these different factors affect the phase transition in VO_2, it becomes of importance to study the structure of the films as the transition takes place at the nanoscale.

Table I: Lattice parameters from the standard VO_2 crystal structures: R-Rutile (tetragonal) and M1-Monoclinic. [15] The order is such as to show the corresponding change in the lattice parameter as the transition takes place.

(R) T > 68°C	4.554 Å a1	4.554 Å a2	2.850 Å a3	$\alpha = 90°$	$\beta = 90°$	$\gamma = 90°$
(M1) T < 68°C	5.383 Å a3	4.526 Å a2	5.753 Å a1	$\beta = 122.8°$	$\alpha = 90°$	$\gamma = 90°$

EXPERIMENTAL DETAILS

Samples for this study were deposited on a silicon wafer with a thermally grown amorphous silicon dioxide layer approximately 380 nm in thickness. A layer of amorphous vanadium oxide (VO_x), close in stoichiometry to VO_2, approximately 50 nm in thickness, was sputtered on top of the thermally grown oxide mentioned above. The sputtering took place by means of DC Magnetron Sputtering in a reactive environment with an oxygen partial pressure.

In order to obtain a crystalline vanadium sample, portions of the wafer were annealed at 450°C under an argon atmosphere for various times. This processing yielded crystalline VO_2 films and isolated particles depending on the annealing time. Though this processing produced crystalline grains with a preferred orientation of the C axis of the tetragonal phase parallel to the plane of the specimen, the grains were randomly oriented in the other directions. [24] Figure 1 shows 4-point probe resistance measurements for some of the films produced. We observe a change in resistance between 2 and 3 orders of magnitude.

The TEM samples for this study were prepared as plan-view wedges using the tripod polishing method. The VO_2 film was left untouched, while the silicon substrate and the amorphous SiO_2 layer were polished down to be electron transparent.

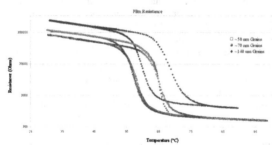

Figure 1: 4-point probe resistance measurements of continuous films annealed at 450°C.

Heating and Imaging the Sample

The data for this study was taken using a Tecnai TF20, 200 kV Field-Emission, High Resolution, Scanning, Analytical TEM. The samples were heated using a Gatan temperature controlled "Cold-Stage." This stage was adequate for this study since the transition temperature for the film was measured near 65°C.

Using the heating stage, the sample was brought from 40°C to 90°C. Reference images were taken at the lower temperature. In order to correct for the drift in the image caused by thermal expansion, the selected area of interest was monitored as heating (and cooling) took place. Both, the stage position (in X and Y) and the focus were corrected, as needed, during changes in temperature.

Selected Area Diffraction, Bright-Field imaging, and Conical Dark-Field Imaging

A selected area aperture was used in order to observe between 50 and 150 grains. In order to determine the appropriate diffraction condition, diffraction ring patterns from the same selected area were taken at 40°C to 90°C and compared against each other. It was imperative to find a ring distinctive and isolated enough between the two phases. Using the smallest objective aperture available, bright-field images were taken for comparison. Contrast in these images was obtained by using the objective aperture to select the main beam in the ring diffraction pattern and blocking as many diffracted beams as possible. This was done to ensure the selected area was the same for the temperatures probed, as well as to observe the difference in contrast for individual grains between the images from their different crystal structures.

Once the appropriate diffraction condition was found, the beam was tilted for dark-field imaging. The Conical Dark-Field (CDF) mode rotates the tilted beam about the central axis thereby exploring all possible diffracted beams of a given tilt angle. To image all these possible angles, the image exposure time is set to be long compared to the conical scan time.

The conditions for the CDF imaging mode were set while the sample was at a temperature of 90°C, and after an adequate diffraction ring was found. The sample was then cooled down by intervals of 10°C until reaching 40°C. Again, image drift and focus were corrected as mentioned above. Time was taken to ensure that the sample was thermally stable enough for an image to be acquired. Once focus was deemed adequate, a CDF image was acquired as mentioned above. Dynamic CDF images were obtained at 90°C, 80°C, 70°C, 60°C, 50°C, and 40°C. The temperature measurements varied no more than ±1°C according to the Gatan temperature controller.

STEM – Convergent Beam Electron Diffraction, Electron Energy Loss Spectroscopy

STEM was employed to probe individual grains as well as the interface between them. The probe was set using a 70 μm aperture, yielding a convergence angle of 9.25 mrad. With this probe, Convergent-Beam Electron Diffraction (CBED) patterns were obtained from individual crystals. Line Scans were set in order to acquire CBED patterns through individual grains, pass across the grain boundary, and into another grain. Crystallinity of individual grains was verified in this manner. Electron Energy-Loss Spectroscopy (EELS) was then used to determine the chemical composition of the material at the grain boundary.

RESULTS AND DISCUSSION

TEM – Selected Area Diffraction, Bright and Conical Dark-Field Images

Figures 2a and 2b show the ring patterns obtained from the selected area shown. These ring patterns were taken at 40°C (2a) and 90°C (2b). For the monoclinic phase there are many more intermediate rings than in the tetragonal phase. There are diffraction rings that are distinct between the monoclinic and tetragonal phases; however, the objective aperture is not able to select them exclusively. The ring labeled "C" in figure 2 was chosen since it was more isolated from the other rings. It was observed that the objective aperture also allowed part of the adjacent diffraction rings. Nonetheless, these two adjacent diffraction rings are observable in both the monoclinic and tetragonal phases, thus differences in contrast arose predominantly due to the chosen ring. It was ensured that the same area was imaged for adequate comparison.

The series of CDF images previously described are shown in figure 2. There are few observable differences while the sample cools from 90°C to 70°C. Some of the grains that were excited at 90°C disappear at 80°C. Still, there are a few that remain excited down to 70°C. The bright contrast observed at temperatures between 90°C to 70°C is attributed to the "adjacent rings" shown in figure 2C. As the transition takes place from the tetragonal to the monoclinic phase, contrast is lost from there rings in order to illuminate diffraction spots in the monoclinic phase. The largest number of particles that switched was observed between 70°C and 60°C (Figure 2e). At this point, increase in contrast from 60°C to 40°C is attributed to the ring in figure 2C. This is consistent with the established transition temperature for the bulk of VO_2. As the temperature kept dropping down to 50°C and 40°C, switching of some particles was still being observed.

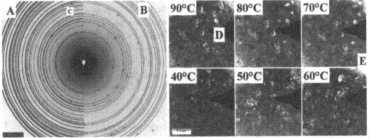

Figure 2: Electron Diffraction Ring Patterns taken at (A) 40°C and (B) at 90°C (overlaid to show calculated d-spacings). The diffraction condition (d-spacing) shown by circle (C) was present in the low-temperature monoclinic phase but missing from the high-temperature tetragonal phase. Dynamic Conical Dark-Field Images obtained at 90°C, 80°C, 70°C, 60°C, 50°C, and 40°C. The temperature measurements varied no more than ±1°C. Differences in contrast between individual grains are attributed to the particles switching out of the high-temperature tetragonal phase into the low-temperature monoclinic phase. A particle (D), used as reference, is circled (red) to aid in tracking changes in the image. The largest number of particles switching was observed between 70°C and 60°C (E).

STEM – Electron Energy Loss Spectroscopy, Convergent Beam Electron Diffraction

Figure 3a shows a representative pair of grains and their interface. EELS line scans were taken from different grains and ensuring data was taken across their grain boundaries. EELS spectra in all cases resembled the spectra expected for VO_2. Though the number of counts varied from grain to grain, the pattern remained the same even while scanning in the grain boundary (Fig 3d), indicating the presence of VO_2 in the grain boundary and not another type of vanadium oxide. High-Resolution images were also taken at the boundaries (Fig 3c). Close examination showed an amorphous grain boundary. The thinnest amorphous grain boundary observed was in the order of 2nm. This grain boundary consisting of amorphous VO_2 may help explain why the resistivity changes by only 2-3 orders of magnitude in thin films as it may behave as a barrier for electron conduction.

CBED patterns were obtained from individual grains, and across the interface. These diffraction patterns did show different crystal orientations for the grains, demonstrating it is possible to obtain structural information from individual particles. Due to the change in structure sets of diffraction spots disappear as the transition takes place (see fig. 2A-B). However, the number of grains with an adequate orientation to view the structural change was very small, thus more work is required to better quantify and understand the phase transition for individual nanoparticles.

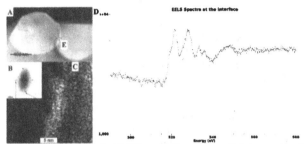

Figure 3: A) STEM Bright-Field image of two VO_2 adjoining grains. The grain boundary was also imaged using High-Resolution TEM (C) showing the different atomic planes of the two grains. The inset (B) is a Fourier-Transform of the HRTEM image showing both lattice parameters. (D) EELS across the grain boundary (E) shows a pattern corresponding to that of VO_2.

CONCLUSIONS AND FUTURE WORK

The TEM allows for the imaging of the sample, as well as the observation of structural changes that take place during the phase transition of VO_2 thin films. Though the grains in the film are randomly oriented, an appropriate diffraction condition was determined through the use of diffraction ring patterns on a selected area of the film. Not all the grains will be oriented to meet that diffraction condition that shows the phase transition. The use of dynamical conical dark-field imaging allows the rough probing of grains that meet the required orientation to show the phase transition. These CDF images taken at cooling intervals of 10°C suggest that the transition temperature for individual particles may vary, though the biggest change was observed between 70°C and 60°C. These TEM techniques are currently being applied to observe the phase transition of individual sub-micron particles.

HRTEM, in conjunction with EELS shows evidence of an amorphous VO_2 as the boundary between grains. The metal to semiconductor transition is known to take place for crystalline VO_2 but not for amorphous VO_2. The difference in resistivity from single VO_2 crystals and thin films may be explained from these grain boundaries.

Current work underway will allow the refinement of these results. The procedure described in this paper helps identify those grains for which the structural phase transition can be observed. CBED can then be employed to further study the structural change in those individual grains.

ACKNOWLEDGMENTS

Special thanks are given to Dr. Kevin Coffey from the University of Central Florida for supplying the samples for this study; to Michael Rawlings for his help in sample preparation; to Dr. Jeffrey Farrer at Brigham Young University for his guidance in the TEM; and finally, to Brigham Young University for their funding and facilities.

REFERENCES

1. F. J. Morin, Phys. Rev. Lett. 3, 34 (1959).
2. M. M. Qazilbash, K. S. Burch, D. Whisler, D. Shrekenhamer, B. G. Chae, H. T. Kim, and D. N. Basov, Phys. Rev. B 74, 205118 (2006).
3. A. Zylbersztejn and N. F. Mott, Phys. Rev. B 11, 4383 (1975).
4. T. Maruyama and Y. Ikuta, Journal of Materials Science 28, 5073 (1993).
5. H. W. Verleur, A. S. Barker, and C. N. Berglund, Phys. Rev. 172, 788 (1968).
6. S. Shin, S. Suga, M. Taniguchi, M. Fujisawa, H. Kanzaki, A. Fujimori, H. Daimon, Y.Ueda, K. Kosuge, and S. Kachi, Phys. Rev. B 41, 4993 (1990).
7. K. A. Khan and M. S. Rahman Khan, Pramana 38, 389 (1992).
8. Moon-Hee Lee and Jun-Seok Cho, Thin Solid Films 365, 5 (2000).
9. Y. Dachuan, X. Niankan, Z. Jingyu, and Z. Xiulin, Journal of Physics D: Applied Physics 29, 1051 (1996).
10. H. J. Schlag and W. Scherber, Thin Solid Films 366, 38 (2000).
11. D. B. McWhan, M. Marezio, J. P. Remeika, and P. D. Dernier, Phys. Rev. B 10, 490 (1974).
12. D. Watanabe, B. Andersson, J. Gjønnes, and O. Terasaki, Acta Crystallographica Section A 30, 772 (1974).
13. D. Kucharczyk and T. Niklewski, Journal of Applied Crystallography 12, 370 (1979).
14. C. Hébert, M. Willinger, D. S. Su, P. Pongratz, P. Schattschneider, and R. Schlögl, European Physical Journal B 28, 407 (2002).
15. Inorganic Crystal Structure Database, http://icsdweb.fiz-karlsruhe.de/index.php.
16. Wang X.J, Li H.D., Fei Y.J., Wang X., Xiong Y.Y., Nie Y.X., and Feng K.A., Applied Surface Science 177, 8 (2001).
17. R. Lopez, T. E. Haynes, L. A. Boatner, L. C. Feldman, and J. R. F. Haglund, Physical Review B. 65, 1 (2002).
18. R. M. B. J. M. Gregg, Applied Physics Letters 71, 3649 (1997).
19. R. M. Wentzcovitch, W. W. Schulz, and P. B. Allen, Phys. Rev. Lett. 72, 3389 (1994).
20. V. Eyert, Annalen der Physik 11, 650 (2002).
21. A. Liebsch, H. Ishida, and G. Bihlmayer, Coulomb correlations and orbital polarization in the metal insulator transition of VO_2 (2003).
22. S. Biermann, A. Poteryaev, A. I. Lichtenstein, and A. Georges, Physical Review Letters 94, 026404 (2005).
23. P. P. Boriskov, A. L. Pergament, A. A. Velichko, G. B. Stefanovich, and N. A. Kuldin, Metal-insulator transition in electric field: A viewpoint from the switching effect (2006).
24. F. Rivera, M.S. Thesis, Brigham Young University (2007). http://contentdm.lib.byu.edu/ETD/image/etd2233.pdf

SYMPOSIUM HH

Mater. Res. Soc. Symp. Proc. Vol. 1184 © 2009 Materials Research Society 1184-HH04-01

The Nature and Characterization of Nanoparticles

Rajiv Kohli
The Aerospace Corporation, Houston, TX 77058-1521, U.S.A.

ABSTRACT

Nanosize particles are of fundamental and practical interest for developing advanced materials and devices and micro and nanostructures. As feature sizes shrink, nanoparticle contamination is also becoming increasingly important to achieve and maintain high product yields. In order to employ appropriate material and product development strategies, or institute preventive assembly and remediation strategies to control nanoparticle contamination, it is essential to understand the nature of nanoparticles and to characterize these particles. Particles in the size range 0.1 nm to 100 nm present unique challenges and opportunities for their imaging and characterization. Critical information for this purpose is the number and size of the particles, their morphology, and their physical and chemical structure. Because of this importance, many advances and new developments have been made in qualitative and quantitative characterization techniques for particles in this size range, including neutron holography, three dimensional atom probe imaging, ultrafast microscopy and crystallography, magnetic resonance force microscopy, and high-resolution x-ray crystallography of non-crystalline structures. It is now possible to completely characterize nanoparticles from 0.1 nm to 100 nm size. A brief review of the nature of nanoparticles is presented and recent developments in selected characterization techniques are described.

INTRODUCTION

In high technology applications across many industrial sectors, component and feature sizes are continually shrinking. The development of new materials for many of these applications involves particle interactions at the nanometer or smaller scale. At the same time, there is increasing realization that further advances in the medical field will require understanding of cellular phenomena at atomic and molecular levels. Characterization of nanometer size particles is essential to understanding their fundamental interactions and their behavior. As an illustration of the importance of characterization at this scale, a new Advanced Materials Laboratory for materials characterization at the near-atomic level has been established at the National Institute of Standards and Technology (NIST) in the United States with stringent environmental controls [1]. This laboratory sets the standard for developing, testing and demonstrating instruments for atomic-level characterization of materials. More recently, a workshop devoted exclusively to the nanostructure problem was held at NIST [2]. The aims of the workshop were to elucidate the challenges associated with accurately determining atomic positions at the nanoscale. This is still a formidable challenge because atom positions must be known to very high precision of the order of 0.01 Å to 0.001 A (1 Å = 10^{-10} m) needed for theoretical calculations of electronic structure and the functional properties of nanostructured materials [3-5]. A major advance in this direction has been to demonstrate measurement of atomic positions at a precision of 0.04-0.06 Å [6-8].

For medically significant molecules, direct observation at the atomic level would reveal the molecular structure of macromolecules in their natural forms, embedded in their natural environments [9]. By combining the chemical specificity of magnetic resonance spectroscopy with the atomic resolution of probe microscopy, the goal of obtaining 3-dimensional images of individual biological molecules could be achieved [10, 11].

In the semiconductor industry, alternative materials are being investigated for integrated circuits that will overcome the fundamental limitations of silicon and silica. The materials of choice are oxides and ionic materials. The electronic properties of these materials can be controlled with nanoscale precision, as shown recently by imaging and manipulating the oxygen vacancies in films of fully oxidized $SrTiO_3$ and of $SrTiO_{3-x}$ [12]. The ability to dope oxide films without introducing impurities is very attractive for commercial semiconductor applications.

Development of new materials and processes also depends increasingly on an understanding of the application of nanosize components, atomic and molecular scale manipulation, and ultrafast interactions. These interactions in atomic, electronic and even nuclear processes occur at femtosecond (10^{-15} seconds) to subzeptosecond (10^{-21} seconds) time scales [13]. Here too electron and probe microscopic techniques are being combined with ultra short laser pulses to track and characterize individual atoms and molecules [13, 14].

From a contamination perspective, nanosize particles can have major impact on the performance of precision and other products. For example, the presence of impurities such as boron and phosphorous in the parts per billion range can result in irrecoverable loss of the entire production lot of semiconductor wafers. Similarly, the presence of low levels of hydrocarbon contaminants in components for oxygen service can cause catastrophic damage of spacecraft due to autoignition. To develop control and remediation strategies for particle contaminants, it is essential to physically and chemically characterize these particles ranging in size from 100 μm to 0.1 nm. These control strategies are also critical to achieving and maintaining high product yields.

In support of these needs, many experimental techniques have been developed and are routinely employed for qualitative and quantitative chemical and physical characterization of particles with near-atomic scale resolution [13, 15]. Examples include electron and probe microscopy, spectroscopic techniques, X-ray and neutron scattering, and diffraction methods. These methods take advantage of the complete range of the structure and properties of materials.

Here we describe recent new developments and applications of selected less common characterization methods that generally apply to bulk samples, but they have the potential to provide quantitative information of nanosize particles at the atomic scale. For example, neutron holography presently requires a sample size of approximately 1 mm^3, but it could be applied to characterize the structure of individual quantum dots arrayed on a substrate [16].

NATURE OF NANOSIZE PARTICLES

In referring to very small particles, the size of the particles can be discussed in terms of various physical phenomena. For example, the interactions of particles much larger than 1 μm diameter is increasingly dominated by gravitational forces, while van der Waals and other forces tend to dominate their interactions below that size. Particles with diameters of 0.3 to 0.7 μm are of the same size as the wavelength of visible light, which is the limit of resolution in optical microscopic observation of particles in that size range. Particles in the size range 20 to 100 nm are referred to as ultrafine, while nanosize particles have diameters smaller than 20 nm. Due to the need to understand aerosol behavior, two additional classes of particle sizes have been defined. Very small particles refer to particles smaller than 5 nm, while molecular size defines particles with diameters smaller than 1 nm [17].

The physical nature of very small particles cannot be thought of in terms of classical surface or volume continua. Rather, the molecules statistically associated with the particle will tend to define their interactions. As Table 1 shows, the number of molecules associated with a particle increases with increasing particle size, but the fraction of the molecules at the surface decreases with increasing particle size. For a particle of 20 nm diameter, the number of molecules at the surface is only about 12%. Consequently, the overall behavior of nanosize particles is governed by the surface and binding energies of the molecules in the particle. These particles are neither solid nor liquid, and do not behave like individual molecules. The particle may be regarded as a complex structure whose behavior depends on the positions of the individual molecules and the combined electronic charge distribution [17].

Table 1. Characteristics of molecules statistically associated with nanosize particles.

Particle Size (nm)	Cross-Sectional Area (10^{-18} m^2)	Mass (10^{-25} kg)	Number of Molecules	% of molecules on the surface
0.5	0.2	0.65	1	-
1.0	0.8	5.2	8	100
2.0	3.2	42	64	90
5.0	20	650	1000	50
10.0	80	5200	8000	25
20.0	320	42000	64000	12

Only simple interactions of nanosize particles are presently understood. Due to the technological importance of very small particles in advanced materials and nanotechnology, advances in kinetic theory, solid state chemistry, quantum mechanics, and aerosol dynamics will make it possible to predict the behavior of very small particles in complex systems in the future. In turn, it will provide the basis for understanding the transport, adhesion and detachment of these particles in real systems, as well as making it possible to design materials with specific properties. One example is the recent synthesis of monodisperse Fe-Pt nanoparticles of controlled size (3–10 nm) and composition to yield chemically and mechanically robust ferromagnetic nanocrystal assemblies for very high areal density magnetic recording (terabits per square inch) [18].

A source of particles is chemically-induced transformation of gaseous matter into particles. These particles can subsequently grow by condensation and coagulation into larger particles at a growth rate that depends on factors such as particle size and concentration [19]. This mechanism is important for assembly processes in which gaseous precursors such as acid gases are present, or for products that employ volatilizable materials, such as motor bearing grease and lubricants. The understanding of gas phase reactions in the formation of particles has advanced considerably, but important chemical and physical mechanisms are still unresolved, particularly for particles in the size range 1–10 nm. As noted earlier, such particles have to be considered as complex structures whose properties and interactions are determined by the number and position of the molecules on the surface. The application of advanced techniques such as 4D ultrafast characterization using femtosecond or even attosecond laser pulses [14, 20] will help resolve many of the gas-particle conversion questions.

CHARACTERIZATION TECHNIQUES

Neutron holography

Holographic techniques have been extended to shorter wavelengths with the development of x-ray and electron holography [21-23]. This has made it possible to image objects with nanometer to atomic resolution. However, both these techniques are still restricted in their applications. Electrons provide high-resolution images, but their strong interactions with condensed matter restrict their use to surface regions. On the other hand, X-rays can penetrate much farther, but their limitation consists of the fact that the penetration depth varies as the square of the atomic number. Therefore, x-ray holography is not very useful for materials with light elements.

Neutrons are not subject to these limitations. Rather than scattering from the electrons in the atoms of the sample, neutrons scatter only from nuclei, which are 100,000 times smaller than the parent atoms. This is an important consideration in the reconstruction of an image of the crystal lattice [24, 25]. Using a highly collimated neutron beam, a neutron holographic image has been recently recorded of lead nuclei in a $Pb_{0.9974}$–$Cd_{0.0026}$ single crystal with atomic-scale resolution [26, 27]. The recorded intensity of the interference pattern between the reference wave and the scattered wave is directly proportional to the emission of γ radiation from the transition of the nucleus in the excited to the ground state. The Pb nuclei act as the object, while the Cd atoms act as highly efficient detectors. This "inside detector" concept" was used to reconstruct the face centered cubic unit cell of twelve lead atoms surrounding the cadmium nucleus. From this reconstructed image, the lattice parameter of the $Pb_{0.9974}$–$Cd_{0.0026}$ single crystal was calculated to be 4.93 Å, in excellent agreement with the value 4.935 Å determined from x-ray diffraction.

Similar results were obtained on a $PdH_{0.78}$ single crystal in which the calculated distance between the hydrogen and palladium nuclei was 2.04±0.06 Å, in very good agreement with the crystallographic value 2.028 Å [28]. The reconstructed hologram in spherical coordinates is shown in Figure 1. The center of the octahedron formed by the Pd spots corresponds to the location of a hydrogen nucleus, making it possible to calculate the distance between the hydrogen and palladium nuclei. The results were recently confirmed using a multi-detector array [29]. This suggests that neutron holography can be applied also in cases where the nuclei are comparatively weakly bound and the zero-point oscillations correspondingly large. These data serve as further quantitative arguments demonstrating the power of the technique for very high precision measurement of the structure.

Elements with atomic number of 45 and higher have isotopes with large cross sections ($\geq$ 5×10 m^2), while about 15 isotopes have such large cross sections for elements with atomic number smaller than 45. Thus, there is a wide variety of materials whose structure can be interrogated with neutron holography using the inside detector concept.

One drawback to the widespread use of neutrons for holographic imaging is the limited intensity of available neutron beam sources. The new high flux spallation neutron sources in the

United States, Japan and Europe will help overcome this limitation. Also, the use of multi-detector arrays can reduce the measuring times from days to hours to record a hologram [30].

Successful demonstration of neutron holography to image crystals with atomic resolution opens up enormous characterization capabilities. For example, since a neutron has a magnetic moment, holography employing polarized neutron beams could contribute to an understanding of the basic magnetic structure of magnetic materials.

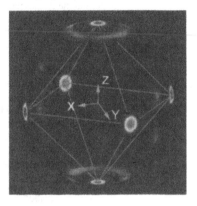

Figure 1. The restored holographic image of a $PdH_{0.78}$ single crystal represented is shown in spherical coordinates in which six Pd atoms surround the hydrogen nucleus [28].

3D atom probe

The three dimensional (3D) atom probe (3DAP) is a quantitative imaging technique that provides atomic scale 3D elemental maps of atoms within a nanoscale volume of a sample [31, 32]. As with scanning tunneling microscopy and high-resolution transmission electron microscopy, a single atom and its neighbors can be imaged. However, 3DAP provides several advantages.

- Elemental analysis can be performed in which each single atom is chemically identified. This makes the technique extremely accurate.
- Because of the way they are evaporated and collected, the positions of the individual atoms within the sample are obtained with subnanometer lateral resolution and the depth resolution of a monolayer. This gives true 3D chemical map of the atoms.
- 3DAP is very precise. From a sample size of 10^6 atoms, the 3DAP can routinely provide compositional information with a standard deviation of ± 0.0001.
- Depending on the setup, the 3DAP can achieve sensitivities of 1 ppm. This can be improved with the next generation of instruments with higher mass resolution and greater number of atoms counted.

In a typical 3DAP instrument, single atoms are field-evaporated from a needle-shaped sample (tip radius ~ 10-70 nm) mounted on a cryogenically-cooled goniometer (Figure 2). This tip is maintained under ultrahigh vacuum condition at 15-70° K during the analysis to ensure reliable composition information is obtained. The atoms are ionized from the surface under a very high electric field and projected toward a position-sensitive detector placed 250-650 mm from the sample. Ionization occurs from the surface of the specimen regularly, which makes it possible to ionize atoms by atomic layer and by atomic order, thereby achieving atomic layer resolution. Atoms are chemically identified by time-of-flight mass spectrometry. The detector gives an accurate measurement of the ion impact positions and masses from the time of flight. The very high magnification of the instrument yields highly accurate (0.2 nm) impact coordinates, from which the atom original positions at the tip surface are derived. Modern 3DAP instruments are equipped with a reflectron energy compensator and can achieve mass resolution higher than 500. This makes to possible to identify most alloying elements in nanocrystalline materials without ambiguity.

For electrically nonconductive samples, an ultrafast laser pulse is used to assist in vaporization of the sample [33]. Femtosecond-pulsed evaporation mode reduces heating of the sample and prevents sample rupture due to thermal cycling stresses by high voltage pulsing. It also avoids the disadvantage of longer duration pulses, such as picosecond laser pulses, which degrade the mass resolution by thermal evaporation and can lead to redistribution/diffusion of the atoms during the pulse, especially if the sample thermal conductivity is low.

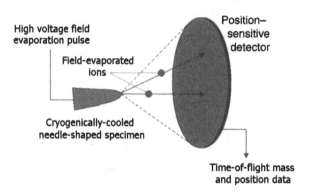

Figure 2. Schematic diagram of a 3D atom probe.

An example of a reconstructed atom map of a single lamellar precipitate in a tool steel is shown in Figure 3 [34]. The precipitate contains mainly vanadium, with molybdenum atoms attached to or embedded in the precipitate. Cobalt atoms are also observed very close to but not generally within the precipitate. The reconstruction map is a section about 7 nm long and 1 nm wide, and extends over 2 atomic layers. This small precipitate could not be observed in high resolution electron microscopy, which shows the distinct advantage of 3DAP for analyzing very small precipitates that are smaller than the typical thin foil thickness of transmission electron microscopy specimens.

The atom probe is particularly well suited to the analysis of the interfaces in thin films. The ability to reveal interface roughness and layer thickness is particularly unique. This rests on its ability to chemically identify and locate all atoms in 3D. The International Technology Roadmap for Semiconductors [35] lists a number of areas of concern such as thin film metrology for which the atom probe is uniquely suited [33].

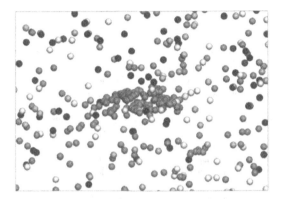

Figure 3. Atom map of a lamellar precipitate in a tool steel containing mainly V atoms (grey) with Mo atoms (light grey) attached to or embedded in the precipitate. Co atoms (black) are observed in close vicinity to the precipitate. The precipitate is about 2.5 nm and a thickness of only two atomic layers [34].

4 Dimensional ultrafast characterization

Ultra short electromagnetic pulses have always been of great interest, largely as a means of investigating and controlling ever faster processes on different time scales: molecular, atomic and electronic. The generation and measurement of ultra short pulses has been recently reviewed [14, 20, 36]. These ultra short pulses can be employed as a synchronized attosecond electron probe, making it possible to control electronic motion with light forces [20]. The shape of the light pulses themselves also can be controlled on a femtosecond timescale, making it possible to maneuver the interacting systems into final states that are hard to reach through classical thermodynamic processes. Such polarization-shaped pulses can be applied to the alignment of molecules in a gas phase or to manipulate larger chiral molecules. The demonstrated experimental tools and techniques of four-dimensional (4D) ultrafast electron diffraction (UED), crystallography (UEC), and microscopy (UEM) open the door to determine complex transient structures and assemblies in different phases. By measuring the duration of atomic processes with an accuracy better than 100 attoseconds, it is now possible to accurately define interactions on this time scale, such as the electron dynamics of hydrated species [37], or the reaction fronts of reactive multilayer foils and nanoscale structures, including gold and graphite films [39-40].

In actual practice, the laser pulses are converted into electron pulses, each electron pulse corresponding to a single image captured by the microscope. By combining the individual images, it is possible to follow the real-time motion of atoms and molecules and study fleeting structures and morphologies, such as atomic structural expansion, nonthermal lattice temperature, ultrafast transients of warping and bulging, dislocation dynamics, twinning, and crystal formation. The processes can be characterized spatially (structural dimensions) and temporally (time dimension). Successful demonstration of these studies opens up a wide range of applications in physical, chemical and biological sciences.

Magnetic resonance force microscopy

The ultimate goal of medical imaging is to image human body cells with atomic resolution. By combining the chemical specificity of magnetic resonance imaging with the atomic level resolution of the scanning probe microscope, it should be possible to produce three-dimensional, in situ images of binding sites in viruses and other features in biological macromolecules [9]. Furthermore the same principle can be applied to image atoms, other macromolecules, defects in solids, and dopants in semiconductors at the atomic scale. The combined technique known as magnetic resonance force microscopy (MFRM) has reached the ultimate in sensitivity by imaging a single electron spin [10].

MRFM is based on detecting the magnetic force between a ferromagnetic tip and the electron spins in a sample. The instrument used to image a single electron spin employs a custom fabricated mass-loaded cantilever 100 nm thick with an attached 150-nm-wide SmCo magnetic tip [10]. An external magnetic field is applied which excites electron spins at a depth of approximately 100 nm in the sample at their resonant frequency. The extremely small force (about 2×10^{-18} Newtons) exerted on the cantilever through the magnetic moment of a single electron spin could be detected at a spatial resolution of 25 nm.

Extending the magnetic resonance force microscopy (MRFM) technique to nuclear spins, necessary for imaging biological molecules, requires at least 1000-fold increase in magnetic moment sensitivity, since nuclear spins interact about 600 times more weakly than electron spins. Very recently, ultrasensitive MFRM has been applied to magnetic resonance imaging to achieve three dimensional imaging of the 1H nuclear spins in a biological specimen with a spatial resolution of 4 nm [11]. This represents an order of magnitude improvement in resolution from a previous imaging experiment in which the spins of 19F nuclei were imaged at 90 nm resolution [41].

Although scanning tunneling microscopes have been used to image individual atoms, they cannot image deeper than one or two atomic layers. As has been demonstrated, the MRFM instrument can detect the electron and nuclear spins of atoms located dozens of atomic layers beneath the surface, making it a three-dimensional investigative tool.

Imaging non-crystalline materials

Many important structures, such as amorphous and disordered materials, biological macromolecules and many polymers, exhibit non-periodic nanostructures that are not accessible to crystallography. To overcome this limitation, new methods have been developed for direct imaging of non-crystalline samples that are applicable with x-ray or electron diffraction [42-48]. By employing coherent radiation, two and three-dimensional imaging of non-crystalline material structures has been demonstrated at a resolution of 8 nm for x-ray and 1 Å for electron diffraction, respectively [46, 47].

The situation for non-crystalline specimens is different from crystalline samples in that the diffraction pattern is continuous rather than concentrating the far-field diffraction pattern into discrete Bragg peaks. This continuous pattern can be sampled on a finer scale from which an image can be reconstructed. The intensity of the diffraction pattern provides a record of the size of the diffraction amplitude. Moreover, there is an unavoidable loss of phase information in the diffraction intensity. To reconstruct an image, requires both the amplitude and the phase of the wave function. This so-called "phase problem" is overcome by oversampling the diffraction pattern [48, 49]. The diffraction intensities are measured in reciprocal space below the Nyquist frequency from which the amplitude of the wave function can be derived. The phase information can be retrieved ab initio from the diffraction intensities through an iterative algorithm [48, 49]. This approach has been successfully applied to image nanocrystals of gold by x-ray diffraction [42, 43] and double–walled carbon nanotubes at 1 Å resolution using nanoarea electron diffraction [44]. A buried Ni pattern could also be successfully imaged nondestructively at 50 nm resolution using soft x-rays [45]. High-resolution imaging can be combined with simultaneous single-molecule fluorescence and Raman spectroscopy [50] to characterize the optical and electronic structure of single-walled carbon nanotubes.

CONCLUSIONS

Recent developments in several techniques for qualitative and quantitative characterization of nanosize-size particles have been described. These less-well known techniques include neutron holography, 3D atom probe, ultrafast microscopy, magnetic resonance force microscopy, and high-resolution x-ray crystallography of non-crystalline structures. Although these techniques are generally employed for characterization of bulk samples, they have the potential to characterize the structure of nanosize particles with precision of 0.01 Å.

REFERENCES

1. NIST Advanced Measurement Laboratory, Fact Sheet, http://www.nist.gov/public affairs/amlbrochure.htm (2009).
2. I. Levin and T. Vanderah, J. Res. Natl. Inst. Stand. Technol. **113**, 321 (2008).
3. S. Van Aert, A. J. den Dekker, and D. Van Dyck, Micron. **35**, 425 (2004).
4. S. Van Aert, D. Van Dyck, and A. J. den Dekker, Opt. Express **14**, 3830 (2006).
5. K. W. Urban, Nature Mater. **8**, 260 (2009).
6. L. Houben, A. Thust, and K. W. Urban, Ultramicroscopy **106**, 200 (2006).
7. S. Bals, S. Van Aert, G. Van Tendeloo, and D. Ávila-Brande, Phys. Rev. Lett. **96**, 096106 (2006).
8. C. L. Jia, S. B. Mi, K. W. Urban, I. Vrejoiu, M. Alexe, and D. Hesse, Nature Mater. **7**, 57 (2008).
9. J. L. Garbini and J. A. Sidles, Program for Achieving Single Nuclear Spin Detection, White Paper-1, Version 2.0c, Quantum System Engineering Group, University of Washington, Seattle, WA (2005).
10. D. Rugar, R. Budakian, H. J. Mamin, and B. W. Chul, Nature **430**, 329 (2004).

11. C. L. Degen, M. Poggio, H. J. Mamin, C. T. Rettner, and D. Rugar, Proc. Natl. Acad. Sci. USA **106**, 1313 (2009).
12. D. A Muller, N. Nakagawa, A. Ohtomo, J. L. Grazul, and H. Y. Hwang, Nature **430**, 657 (2004).
13. R. Kohli, in *Particles on Surfaces 9: Detection, Adhesion and Removal*, edited by K. L. Mittal, VSP, Utrecht, 2006) p. 3.
14. A. H. Zewail, Annu. Rev. Phys. Chem. **57**, 65 (2006).
15. R. Kohli, in *Particles on Surfaces 8: Detection, Adhesion and Removal*, edited by K. L. Mittal, VSP, Utrecht, 2003) p. 3.
16. G. Krexner (private communication).
17. O. Preining, in *Developments in Surface Contamination and Cleaning. Fundamentals and Applied Aspects*, edited by R. Kohli and K. L. Mittal (William Andrew, Norwich, NY, 2008), p. 3.
18. S. Sun, C. B. Murray, D. Weller, L. Folks, and A. Moser, Science **287**, 1989 (2000).
19. S. K. Friedlander, *Smoke, Dust and Haze: Fundamentals of Aerosol Dynamics*, 2nd Edition (Oxford University Press, New York, NY, 2000).
20. F. Krausz and M. Ivanov, Rev. Mod. Phys. **81**, 163 (2009).
21. M. Tegze, G. Faigel, S. Marchesini, M. Belakhovsky, and A. I. Chumakov, Phys. Rev. Lett. **82**, 4847 (1999).
22. A. Tonomura, Editor, *Electron Holography*, 2nd Edition (Springer, New York, 1999).
23. P. A. Midgley and R. E. Dunin-Borkowski, Nature Mater. **8**, 271 (2009).
24. L. Cser, G. Krexner, and Gy. Török, Europhys. Lett. **54**, 747 (2001).
25. B. Sur, R. B. Rogge, R. P. Hammond, V. N. P. Anghel, and J. Katsaras, Nature **414**, 525 (2001).
26. L. Cser, Gy. Török, G. Krexner, I. Sharkov, and B. Faragó, Phys. Rev. Lett. **89**, 175504 (2002).
27. L. Cser, G. Krexner, M. Markó, I. Sharkov, and Gy. Török, Phys. Rev. Lett. **97**, 255501 (2006).
28. L. Cser, Gy. Török, G. Krexner, M. Prem, and I. Sharkov, Appl. Phys. Lett. **85**, 1149 (2004).
29. H. Kouichi, K. Ohoyama, S. Orimo, Y. Nakamori, H. Takahashi, and K. Shibata, Jap. J. Appl. Phys. **47**, 2291 (2008).
30. L. Cser, G. Krexner, M. Markó, M. Prem, I. Sharkov, and Gy. Török, Physica B. **385-386**, 1197 (2006).
31. M. K. Miller, *Atom Probe Tomography, Analysis at the Atomic Level* (Kluver Academic/Plenum, New York, NY, 2000).
32. D. N. Seidman, Annu. Rev. Mater. Res. **37**, 127 (2007).
33. B. Gault, F. Vurpillot, A. Vella, M. Gilbert, A. Menand, D. Blavette, and B. Deconihout, Rev. Sci. Instrum. **77**, 043705 (2006). See also http://www.cameca.fr.
34. M. Niederkofler and M. Leisch, Appl. Surf. Sci. **235**, 132 (2004).
35. International Technology Roadmap for Semiconductors, 2008 Edition, International Sematech, Austin, TX (2008).
36. C. Altucci and D. Paparo, "Elucidating the Fundamental Interactions of Very Small Particles: Ultrafast Science," in *Developments in Surface Contamination and Cleaning. Fundamentals and Applied Aspects*, edited by R. Kohli and K. L. Mittal (William Andrew, Norwich, NY, 2008), p. 25.

37. A. E. Bragg, J. R. R. Verlet, A. Kammrath, O. Cheshnovsky, and D. M. Neumark, Science **306**, 669 (2004).
38. J. S. Kim, T. LaGrange, B. W. Reed, M. L. Taheri, M. R. Armstrong, W. E. King, N. D. Browning, and G. H. Campbell, Science **321**, 1472 (2008).
39. D.-S. Yang, C. S. Lao, and A. H. Zewail, Science **321**, 1660 (2008).
40. B. Barwick, H. S. Park, O.-H. Kwon, J. S. Baskin, and A. H. Zewail, Science **322**, 1227 (2008).
41. H. J. Mamin, M. Poggio, C. L. Degen, and D. Rugar, Nature Nanotechnol. **2**, 301 (2007).
42. J. Miao, P. Charalambous, J. Kirz1, and D. Sayre, Nature **400**, 342 (1999).
43. I. K. Robinson, I. A. Vartanyants, G. J. Williams, M. A. Pfeifer, and J. A. Pitney, Phys. Rev. Lett. **87**, 195505 (2001).
44. J. M. Zuo, I. A. Vartanyants, M. Gao, R. Zhang, and L. S. Nagahara, Science **300**, 1419 (2003).
45. J. Miao, T. Ishikawa, B. Johnson, E. H. Anderson, B. Lai, and K. O. Hodgson, Phys. Rev. Lett. **89**, 088303 (2002).
46. Q. Shen, I. Bazarov, and P. Thibault, J. Synchrotron Rad. **11**, 432 (2004).
47. D. Shapiro, P. Thibault, T. Beetz, V. Elser, M. Howells, C. Jacobsen, J. Kirz, E. Lima, H. Miao, A. M. Neiman, and D. Sayre, Proc. Natl. Acad. Sci. USA **102**, 15343 (2005).
48. W. J. Huang, J. M. Zuo, B. Jiang, K. W. Kwon, and M. Shim, Nature Physics **5**, 129 (2009).
49. J. Miao, D. Sayre, and H. N. Chapman, J. Opt. Soc. Am. A **15**, 1662 (998).
50. A. Hartschuh, H. N. Pedrosa, L. Novotny, and T. D. Krauss, Science **301**, 1354 (2003).

Mater. Res. Soc. Symp. Proc. Vol. 1184 © 2009 Materials Research Society   1184-HH01-05

Advanced TEM Characterization for Catalyst Nanoparticles Using Local Adaptive Threshold (LAT) Image Processing

Petra Bele[1] and Ulrich Stimming[1,2,3]
[1] Technische Universität München TUM, Department of Physics E19, James-Franck-Strasse 1, 85748 Garching, Germany
[2] nanotum, Technische Universität München TUM, , James-Franck-Strasse 1, 85748 Garching, Germany
[3] Zentrum für Angewandte Energieforschung Bayern ZAE, Division 1, Walther-Meissner-Strasse 1, 85748 Garching, Germany

ABSTRACT

Metallic and non-metallic nanoparticles, usually supported on non-metallic substrates have attracted much interest concerning their application in the field of electrocatalysis. To characterize catalysts with respect to size, morphology, structure and composition (alloys or core-shell) of nanoparticles and their associated electrocatalytic activity, transmission electron microscopy (TEM) is the state of the art method. This investigation shows the advantages of advanced image processing using the local adaptive threshold (LAT) routine.

INTRODUCTION

For a thorough structural TEM characterization of catalysts, represented by small nanoparticles on a matrix, one has to deal with obstacles due to image detection and image processing [1]; in the case of image detection with:

i.   variation of image contrast due to local thickness changes of the support material,
ii.   intensity variation of similar nanoparticles based on diffraction contrast,
iii.   weak signal-to noise-ratio due to the difficulty to distinguish particles in the sub-nanometer scale from the matrix, and
iv.   overlapping of different particles when imaged in projection.

In order to overcome these problems, computer image processing methods offer a major advantage in the data evaluation process. However, computer-assisted analysis techniques of TEM images dealing with nanoscaled or even sub-nm particles have their own difficulties arising from the applied image processing routines itself [2-3]. Therefore, a function is needed to obtain the image segmentation, which involves the classification of each image pixel to one of the image parts, either object or background.

EXPERIMENTAL DETAILS

To strive for the most objective results an advanced computerized image processing routine is introduced to evaluate particle size and size distribution. A more detailed description can be found in [4]. The key for the final determination of particle diameter is to use the so-called local adaptive threshold (LAT) routine instead of the standard global threshold routine before particle picking. By using just a global threshold, one typically has to deal with loosing too much of the desired region or getting too many extraneous background pixels resulting in an under- or overestimation of the desired region. In addition, illumination changes across the image can occur, causing brighter and darker parts not correlated to the real objects in the image. LAT typically takes a grey-scale image as input and outputs a binary image representing the segmentation assuming that smaller sub-image regions are more likely to have approximately uniform illumination compared to the complete image. Applying the LAT is the key for an exact determination, even for particles in the sub-nanometer scale, leading to a higher degree of accuracy concerning the complete data analysis process. This is demonstrated by a study using different catalysts on different support materials combining TEM data with results received from other electrochemical experiments.

DISCUSSION

Here the advantages of the image processing routine using LAT is presented. The first example will demonstrate the achievement resulting in unbiased particle determination. The second experiment reveals the advantage against the still quite common 'by hand and eye' method and the last example will demonstrate the possibility to correlate results received from TEM characterization and electrochemical experiments in order to calculate the electrochemically active area of the catalyst nanoparticles.

Test Imaging Processing Routine

The first experiment was done to evaluate the quality of the advanced image processing routine. Therefore, a model system (10 wt.% Pt on Glassy Carbon) was used. TEM sample preparation was made by carefully removing the topmost layer of the Pt/GC electrode with a scalpel. Images were taken with a JEOL 2010, working at 120kV with a LaB_6 cathode.

In Figure 1 a typical TEM image is depicted showing raw data to demonstrate the previous discussed problems. The corresponding histogram reveals the statistically relevant nanoparticle determination. From the data it is easy to obtain the following information:

    i.   particle size distribution, depicted always in a histogram with different possible classifications;

    ii.   the particle diameter with 3 defined values. d_{mean} is the statistical average over all picked particles, d_{median} defines the value where half of the number of particles lie above this value and d_{Fit} is the value received by fitting the histogram with a lognormal distribution function;

    iii.   information about the particle distribution on defined support area and

    iv.   one can calculate the ratio surface area$_{(particles)}$ to surface area$_{(support)}$.

 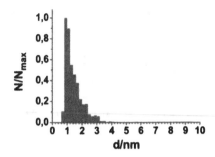

Figure 1. Typical bright field TEM image (raw data) of 10wt.% Pt on Glassy Carbon (left) and corresponding particle size distribution (classification 0.2) according to image processing routine using LAT (right) [4].

Comparing Data Evaluation Between 'By Hand and Eye' Method and Advanced Image Processing

In order to show the need for computerized data evaluation a comparison is made between 'by hand and eye' method and image processing routine with LAT. For both evaluations previously taken STEM images of Pt_3Co NPs on carbon black were used [5].

In this figure the advantage of the advanced image processing over previously presented results is obvious. For the manual particle picking only a classification of 0.5 is possible and resulting in 500 picked particles. The advanced image processing is able to distinguish different classification (here 0.5 and 0.2) and the number of picked particles is a factor of 3 higher, even though the same images were used. By fitting the histograms with a log-normal distribution function the advantage of the advanced image processing is demonstrated.

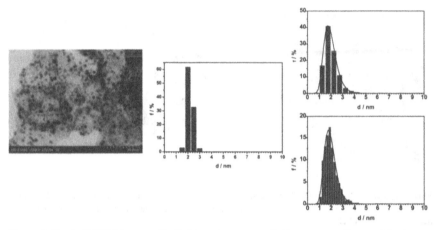

Figure 2. Typical STEM image of Pt_3Co/C catalyst system (left); corresponding particle size distribution derived by manual particle picking (middle) and related particle size distributions for 2 different classifications (right) [6].

Electrochemically Active Area - Calculation and Comparison of TEM Results with Cu-UPD Measurements

The last example demonstrates the possibility to compare results from ex-situ TEM investigations with in-situ electrochemical measurements. The underpotential deposition of copper (Cu-upd) is well established [7, 8] as a method to determine the electrochemically active areas of catalyst nanoparticles. Therefore, a reference system of 10wt.% Pt on carbon black was used.

The electrochemically active surface areas were determined by Cu-UPD and, for comparison, H-UPD technique. From the electrochemical data in figure 2, the electrochemically active area S was determined by Cu-UPD. For the 10wt.% Pt/C catalyst system a value of $S = 72 \text{ m}^2\text{g}^{-1}_{Pt}$ was obtained.

The calculation of the theoretical active area S_{theo} was made using 1522 particles derived by TEM imaging and advanced image processing and assuming spherical nanoparticles. This results in a total area $A_{particles}$ of $2\times10^{-14} \text{ m}^2$ and a volume of $V_{particles}$ of $1\times10^{-23} \text{ m}^3$; with a density of Pt of 21450 kgm^3 the theoretical active area S_{theo} can be estimated to $93 \text{ m}^2\text{g}^{-1}_{Pt}$.

The difference of approximately 13% between both values can be explained by:

i. neglecting agglomerates in TEM evaluation;
ii. nanoparticles as well as agglomerates are detected by electrochemical techniques;
iii. Pt particles which are visible in TEM but are electrochemically inactive and
iv. neglecting the carbon support morphology.

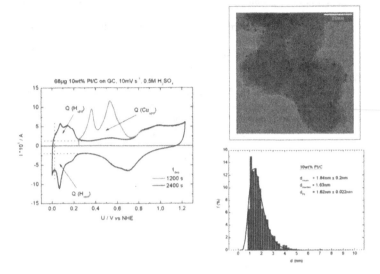

Figure 3. Base voltammogram of hydrogen adsorption and desorption and Cu-UPD charge (left); typical bright field TEM image and corresponding particle size distribution (right).

CONCLUSIONS

The advanced image processing for TEM images using local adaptive threshold leads to a fast, reliable and statistically meaningful characterization of nanoparticles in terms of size distribution, diameter even in the sub-nanometer region, distribution on defined support area, the ratio area$_{particles}$ to area$_{support}$ and the possibility to calculate and compare the catalytically active area. These results built a fundamental base for future detailed investigations on structure/reactivity-relationship for nano-structured electrocatalysts.

ACKNOWLEDGMENTS

Petra Bele would like to thank the Deutsche Forschungsgemeinschaft DFG for the grant BE4154/3-1 AOBJ: 565443.

REFERENCES

1.   L. Reimer, "*Transmission Electron Microscopy - Physics of Image Formation and Microanalysis*" (Springer-Verlag, Germany, 1997).

2.   M. T. Reetz, M. Maase, T. Schilling and B. Tesche, *J. Phys. Chem. B* **104**, 8779-8781 (2000).
3.   J. C. Russ, *"The Image Processing Handbook"*, 2nd Edition, (CNR Press, Boca Raton, USA, 1995).
4.   P. Bele, F. Jäger and U. Stimming, *Microscopy and Analysis*, **21(6) EU**, S5-S7 (2007).
5.   H. Yano, M. Kataoka, H. Yamashita, U. Uchida and M. Watanabe, *Langmuir* **23**, 6438-6445 (2007).
6.   P. Bele, U. Stimming, H. Yano, H. Uchida and M. Watanabe, *Imaging and Microscopy* (2009, submitted).
7.   S. Trasatti and O. A. Petrii, *Pure & Appl. Chem.* **63**, 711 (1991).
8.   E. Lamy-Pitara and J. Barbier, *Appl. Catal. A* **149**, 49 , (1997).

Mater. Res. Soc. Symp. Proc. Vol. 1184 © 2009 Materials Research Society   1184-HH01-06

Quantitative Scanning Transmission Electron Microscopy for the Measurement of Thicknesses and Volumes of Individual Nanoparticles

Helge Heinrich[1,2,3], Biao Yuan[1,3], Haritha Nukala[1,3], and Bo Yao[1,3]
[1]Advanced Materials Processing and Analysis Center (AMPAC), University of Central Florida, Orlando, FL 32816, U.S.A.
[2]Department of Physics, University of Central Florida, Orlando, FL 32816, U.S.A.
[3]Department of Mechanical, Materials, & Aerospace Engineering, University of Central Florida, Orlando, FL 32816, U.S.A.

ABSTRACT

In Scanning Transmission Electron Microscopy (STEM) the High-Angle Annular Dark-Field (HAADF) signal increases with atomic number and sample thickness, while dynamic scattering effects and sample orientation have little influence on the contrast. The sensitivity of the HAADF detector for a FEI F30 transmission electron microscope has been calibrated. Additionally, a nearly linear relationship of the HAADF signal with the incident electron current is confirmed. Cross sections of multilayered samples for contrast calibration were obtained by focused ion-beam (FIB) preparation. These cross sections contained several layers with known composition. A database with several pure elements and compounds has been compiled, containing experimental data on the fraction of electrons scattered onto the HAADF detector for each nanometer of sample thickness. Contrast simulations are based on the multi-slice formalism and confirm the differences in HAADF-scattering contrast for the elements and compounds. TEM offers high lateral resolution, but contains little or no information on the thickness of samples. Thickness maps in energy-filtered transmission electron microscopy, convergent-beam electron diffraction and tilt series are so far the only methods to determine thicknesses of particles in a transmission electron microscope. We show that the calibrated HAADF contrast can be used to determine the thicknesses of individual nanoparticles deposited on carbon films. With this information the volumes of nanoparticles with known composition were determined.

INTRODUCTION

Transmission electron microscopy (TEM) is a very powerful tool with a high lateral resolution of about 2 Å, which allows to image the individual atomic columns giving an insight into the atomic structure of materials. To get a comprehensive picture of a material it is often necessary to measure all three dimensions of a sample. The measurement of the sample thickness is very important to give a full three-dimensional description of each nanoparticle. Though TEM has a high lateral resolution, it is extremely difficult to determine the third dimension (sample thickness). The Atomic Force Microscope (AFM) provides a three-dimensional surface profile; it mainly uses the height of its probe tip to measure the surface topography. At high edges of a surface the AFM image yields incorrect height information as the AFM tip has a finite radius of curvature. Atomic Force Microscopy and TEM are complementary to each other: AFM has a better height resolution than lateral resolution while TEM offers a good lateral resolution but a poor height resolution. In this paper a direct method for the thickness measurement of samples in TEM from a single electron micrograph is presented.

119

EXPERIMENT

A Tecnai F30 from FEI was operated at 300 kV. In the STEM mode a Fischione High-Angle Annular Dark-Field (HAADF) detector with a contrast/brightness setting of 12.5%, and 46.875% and at a camera length of 80 mm was consistently used throughout this study. With a Gatan CCD camera the linearity of the HAADF detector was tested. Samples for TEM were prepared by focused ion beam milling using a FEI 200 TEM system. The calibration for the quantification of the atomic number-contrast (Z-contrast) method for the determination of the sample thickness can be done by wedge methods, using e.g. Ag_2Al platelets in an Al matrix [1] or a pure Silicon wafer cleaved parallel to {111} planes [2], or by calibrations using convergent-beam electron diffraction patterns [3] at specific sample locations. With locally known thicknesses the HAADF intensity was normalized into counts per nanometer thickness. Multilayer systems provided by TriQuint Semiconductors in Apopka (FL) were used to compare HAADF intensities of neighboring layers, assuming no thickness change across the interface. This yields relative intensities of layers for several single elements and compounds.

Simulations in C# .NET 3.0 for the HAADF-STEM intensity of aluminum was performed with procedures originally developed by Erni et al. [4]. In Fig. 1 the resulting absolute intensity (the fraction of electrons scattered into a specific angular interval) is plotted against the number of Al layers in [100] orientation: The conditions used were: Acceleration Voltage: 300 KV; Spherical Aberration: 1.2 mm; Defocus: 57 nm; Condenser Aperture Radius: 7 mrad; Sampling Rate: 0.004 nm/pixel; Detector angles: 56 mrad – 246 mrad. The resulting intensity increases as counts/nm with increasing simulated thickness.

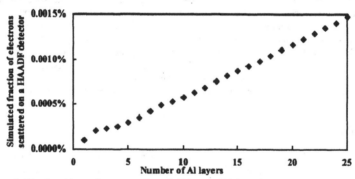

Figure 1. Simulated intensity vs. number of Al layers (thickness)

The HAADF intensities are calibrated by measuring the HAADF detector intensities when the electron beam is directly (in imaging mode) scanned across the detector. This procedure is repeated for different electron beam intensities, which are modified by changing spot size, condenser aperture size, gun-lens setting, and extraction voltage in the transmission electron microscope. For each electron beam setting the intensity measured by the HAADF detector is correlated with the intensity per second measured by the CCD camera. The HAADF detector shows some variability (about 10%) of its sensitivity depending on where the electron beam hits the detector (see insert in fig. 2). At the specific contrast-brightness setting used the HAADF detector intensity scales with the linear CCD signal according to $I(HAADF) \sim I(CCD)^{1.08}$.

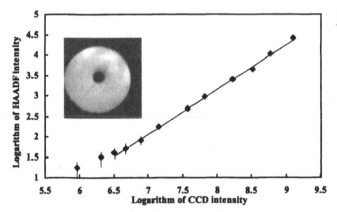

Figure 2. Intensity of the HAADF detector after background subtraction as a function of the CCD camera signal for different electron probes. The insert shows areas of different sensitivity on the HAADF detector using the image mode (not the diffraction mode) in STEM.

The experimental contrast calibration of intensity vs. atomic number is shown in Fig. 3. With the average intensity on the detector for a specific spot size and condenser aperture size, the relative intensities (counts per nanometer) were transformed into the fraction of electrons scattered onto the HAADF detector for each nanometer of sample thickness for the respective material. Furthermore, by dividing these data by the atomic density of a material, we get can the value for σ, the interaction coefficient of the scattering cross-section for the electrons scattered by an atom to the corresponding scattering angles of the HAADF detector.

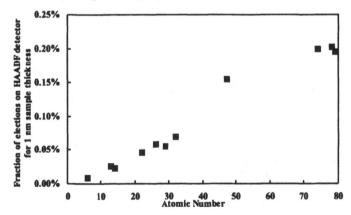

Figure 3. Fraction of electrons scattered onto the HAADF-STEM detector per nanometer of sample thickness for different pure elements.

The data from Fig. 3 confirm that the interaction cross section increases with an increase in the atomic number. However, the scattering signal does not follow pure Rutherford scattering proportional to Z^2, as the scattering signal for these scattering angles is influenced by the electron contribution of each atom.

For contrast calibrations of different elements multilayered samples were used. One example shown in Fig. 4 with six layers of [Pt(28nm)/Fe(22nm)] on top of a SiO_2 coated a Si substrate. With HAADF-STEM the Pt and Fe contrast was calibrated using these as-deposited Pt-Fe multilayers where interdiffusion is not occurring (Fig. 4). The thickness of the sample prepared by focused ion beam processing was determined by convergent-beam electron diffraction of the Si substrate.

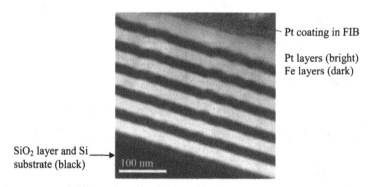

Pt coating in FIB

Pt layers (bright)
Fe layers (dark)

SiO_2 layer and Si
substrate (black)

100 nm

Figure 4. HAADF STEM micrograph of a multilayer system on Si and SiO_2 with six Pt (28 nm) and six Fe (22 nm) layers.

RESULTS

Au-Fe nanoparticles were studied with conventional TEM and HAADF STEM (Fig. 5). The local thickness of nanoparticles (as shown in Figure 6) was measured yielding also the volume of the nanoparticles. The intensity/thickness ratio for the nanoparticles was calculated. The right part of Fig. 6 shows the intensity across nanoparticles along the arrow in the calibrated micrograph of the $Au_{0.5}Fe_{0.5}$ nanoparticles in Fig. 6 left.

The total intensity for each particular nanoparticle has been measured by integrating the thickness over the area of the nanoparticle and by multiplying the result with 0.066 nm^2 (since 1 pixel at the magnification used has a lateral dimension of 0.257 nm). This yields a volume of (170 ± 20) nm^3 for the nanoparticle in the top part of the linescan (Fig. 6, left). For the second nanoparticle of this linescan, a volume of (260 ± 20) nm^3 is measured. Similarly, volumes of other nanoparticles were determined. The average particle volume determined from all nanoparticles shown in Fig. 6 is (80 ± 20) nm^3. This corresponds to a typical particle diameter of a little more than 5 nm.

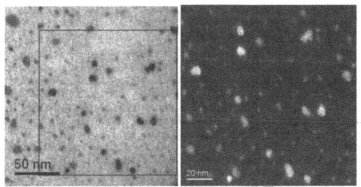

Figure 5. Conventional TEM (left) and STEM micrographs (right) of $Au_{0.5}Fe_{0.5}$ nanoparticles, a sample provided by Dr. Roldan of UCF's Physics Department. The contrast of the STEM micrograph of the $Au_{0.5}Fe_{0.5}$ nanoparticles yields data on the thicknesses and volumes of individual nanoparticles.

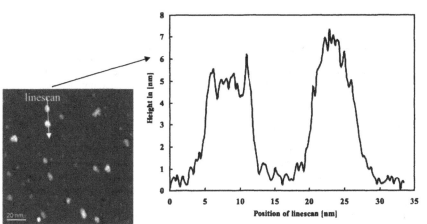

Figure 6. Left: Calibrated HAADF-STEM thickness map obtained from Fig. 4 (right). The heights of two nanoparticles along the line in Fig. 6 (left) are shown on the right.

The HAADF-STEM measurements show that these Au-Fe nanoparticles have volumes of up to 300 nm^3. Nanoparticles with volumes as small as 30 nm^3 were clearly identified and distinguished from HAADF signal variations caused by different C film thicknesses or polymeric residues at the surfaces of the Au-Fe nanoparticles. The measurement of the thickness of the nanoparticles is important as the change in the thickness as well as the overall size of nanoparticles may bring about a change in the catalytic properties [5], in luminescence-center-mediated excitation [6], or other applications where they have very important functions.

CONCLUSIONS

For thin samples the HAADF contrast should follow a simple approach described by Heidenreich [7] with $I \sim t\rho Z^{\alpha}$, where t is the sample thickness, ρ is the atomic density, Z is the atomic number, and α is a parameter between 1 and 2. However, with incorporation of the atomic densities the data from Fig. 3 reveal that this simple approach cannot be used for all elements. There is no simple relationship between atomic number and HAADF-STEM contrast which can be described by only one parameter α. Thus, quantitative HAADF-STEM calculations of thicknesses or compositions require intensity calibration for each element or compound separately. The simulated results of the scattering amplitudes are consistent with the experimental results. The HAADF STEM signal shows only limited dynamic scattering effects and orientation effects and is mainly determined by incoherent scattering. The volumes of nanoparticles were determined for samples with known composition from a single HAADF-STEM micrograph without tilting of the sample. This method is especially useful for the measurement of volumes of nanoparticles with non-spherical shape. For binary alloys with known TEM sample thickness, the composition can be determined from a single HAADF-STEM micrograph.

ACKNOWLEDGMENTS

The authors acknowledge Professor Beatriz Roldan in the Department of Physics at the University of Central Florida for proving samples of Au-Fe nanoparticles. The authors thank TriQuint Semiconductors in Apopka, FL and the UCF-OORC for financial support, and Drs. Gernot Fattinger, Taeho Kook, and Robert Aigner from TriQuint for calibration samples.

REFERENCES

1. H. Heinrich, "High-Resolution Transmission Electron Microscopy for Nanocharacterization", in: *"Functional Nanostructures. Processing, Characterization, and Applications"*, Edited by S. Seal, in series "Nanostructure Science and Technology", series editor: D.J. Lockwood, for a book entitled "Functional Nanostructures" (Editor: S. Seal) Springer Science and Business Media, LLC, 414-503 (2007).
2. Haritha Nukala, MS Thesis, University of Central Florida, Summer 2008.
3. M. Tanaka, M. Terauchi, T. Kaneyama, "Convergent-Beam Electron Diffraction II" (JEOL, Tokio, 1988).
4. R. Erni, H. Heinrich, G. Kostorz, "Quantitative characterisation of chemical inhomogeneities in Al-Ag using high-resolution Z-contrast STEM", *Ultramicroscopy* **94**, 125-133 (2003).
5. J.R.Croy, S.Mostafa, J.Liu, Y.Sohn, H.Heinrich, B.R.Cuenya, *Catal Lett*, **199** 209 (2007).
6. O.Savychyn, F.R.Ruhge, P.G.Kik, R.M.Todi, K.R.Coffey, H.Nukala, H.Heinrich, *Phys Rev B* **76**, 195419 (2007).
7. R. D. Heidenreich, *"Fundamentals of Transmission Electron Microscopy"*, Wiley, New York, 1964.

Mater. Res. Soc. Symp. Proc. Vol. 1184   1184-HH01-08

Electron Microscope Study of Strain in InGaN Quantum Wells in GaN Nanowires*

R. H. Geiss, K. A. Bertness, A. Roshko, and D. T. Read

National Institute of Standards and Technology, 325 Broadway, Boulder, CO 80305, U.S.A.

ABSTRACT

Strains in GaN nanowires with InGaN quantum wells (QW) were measured from transmission electron microscope (TEM) images. The nanowires, all from a single growth run, are single crystals of the wurtzite structure that grow along the <0001> direction, and are approximately 1000 nm long and 60 nm to 130 nm wide with hexagonal cross-sections. The In concentration in the QWs ranges from 12 to 15 at %, as determined by energy dispersive spectroscopy in both the transmission and scanning electron microscopes. Fourier transform (FT) analyses of <0002> and <$1\bar{1}00$> lattice images of the QW region show a 4 to 10 % increase of the c-axis lattice spacing, across the full specimen width, and essentially no change in the a-axis value. The magnitude of the changes in the c-axis lattice spacing far exceeds values that would be expected by using a linear Vegard's law for GaN – InN with the measured In concentration. Therefore the increases are considered to represent tensile strains in the <0001> direction. Visual representations of the location and extent of the strained regions were produced by constructing inverse FT (IFT) images from selected regions in the FT covering the range of c-axis lattice parameters in and near the QW. The present strain values for InGaN QW in nanowires are larger than any found in the literature to date for other forms of $In_xGa_{1-x}N$ (QW)/GaN.

INTRODUCTION

The interesting and technologically promising optoelectronic properties of InGaN quantum wells on GaN have been amply documented [1-3]. The role of strain in controlling the optoelectronic behavior of InGaN-GaN structures has also been well documented [4,5]. Here we report measurements of strain in and around InGaN quantum wells in GaN nanowires made by Bertness et al. [6]. This geometry is of interest both for its technological possibilities and for the insight it provides into the behavior of the InGaN-GaN material system.

We describe and apply a novel computer-interactive approach to measuring and imaging local strains using high resolution TEM imaging, based on a Fourier transform (FT) technique. A number of FT methods for measuring strain by use of high resolution lattice imaging have been reported [7-10]. We suggest that our approach is much more physically intuitive and easy to use than the previously reported methods.

*Contribution of the U.S. National Institute of Standards and Technology. Not subject to copyright in the U.S.

PROCEDURES

Experimental

In$_x$Ga$_{1-x}$N (QW)/GaN nanowires were grown on a (111) Si substrate with molecular beam epitaxy (MBE), with elemental In and Ga and an RF-plasma N$_2$ source [6]. The substrate temperature was 820 ± 10 °C for the GaN nanowire and 530 ± 20 °C for the In$_x$Ga$_{1-x}$N quantum wells (QW). The crystallographic structure of the nanowires is wurtzite, with a = 0.3189 nm and c = 0.51856 nm as determined by x-ray diffraction [11]. All of the nanowires investigated were from a single growth run. Approximate local chemical compositions of the QW were obtained by use of energy dispersive spectroscopy (EDS) instruments in both the TEM and the scanning electron microscope (SEM); the In concentration in the QW was 14 ± 2 at %.

The growth axis of the nanowires was normal to the (111) surface of the Si substrate [6]. Transmission electron microscopy samples were prepared by dragging a C-mesh TEM specimen support along the surface. The nanowires had hexagonal cross sections with projected widths of 60 nm to 130 nm. The InGaN quantum well was typically about 80 nm from the end of the nanowire and about 10 nm thick. A TEM image of a typical nanowire including a QW is shown in figure 1.

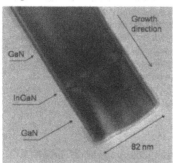

Figure 1. TEM image of the end of a typical GaN nanowire showing the InGaN quantum well near one end. This image is typical of the structure that was imaged at high magnification in the TEM for strain analysis.

TEM imaging was done in two microscopes having very different electron-optical properties [12]. One was a JEOL 2000 FX with coefficient of spherical aberration C_s = 2.3 mm at 200 kV; samples were oriented by use of a single-tilt sample holder. The point resolution at the Scherzer defocus was 0.24 nm. The other microscope was a FEI Titan 80-300 with a C_s-corrected objective; a double-tilt holder was used to orient the specimens. The Scherzer point resolution of the microscope was approximately 0.1 nm at 300 kV.

The growth axis of the nanowires was the c-axis, <0001>. With the single-tilt holder most nanowires could be tilted to an orientation with diffraction vector g = ±0002. In some nanowires, an orientation with g = ±1$\bar{1}$00 could also be achieved, but we were not able to obtain the [11$\bar{2}$0] zone axis orientation. Consequently only lattice fringe images could be obtained. With the double-tilt holder, available in the Titan, we were able to orient nanowires along the [11$\bar{2}$0] zone axis and obtain lattice structure, or dot, images.

The images were recorded in both microscopes by use of charge-coupled device cameras located below the viewing chamber. In the JEOL TEM the images were recorded at a resolution of approximately 1k x 1k pixels, and in the Titan, 2k x 2k pixels. The original magnification for most images was around 500,000, and exposures of 2 s were typical.

Manual-interactive analysis of TEM images

Fourier transforms (FT) of the high resolution TEM images were obtained by use of either Digital Micrograph, a commercial program, or ImageJ, the PC version of the freeware program NIH Image [12]. FT images from unstrained GaN and graphite were used to calibrate the size of the spot corresponding to the unstrained or reference lattice spacing in the FT image, as well as the spatial frequency associated with the centroid location. The average diameter of the FT spot in a number of such reference images from both microscopes was 4 pixels, so this value was taken as the nominal FT spot size for an unstrained lattice. A three-beam, g = ± 0002, lattice image from an InGaN QW and the associated FT image are shown in figure 2. Careful examination of the FT shows that the (0002) spot is elongated in the axial direction of the nanowire. The spot extends 12 to 15 pixels, implying the presence in the real-space image of a distribution of (0002) lattice spacings, rather than a single uniform lattice parameter.

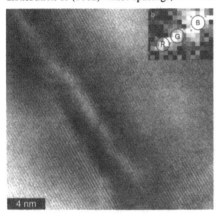

Figure 2. (a) High magnification lattice fringe image, enhanced by Fourier filtering, of an InGaN QW, obtained with three-beam imaging conditions in the JEOL 2000FX. The dark regions are thought to be caused by a variable strain field at the edges of the QW due to non-uniform In concentration. (b, inset) The (0002) spot in the FT of the TEM image, inset, is significantly smeared out from the (0002) center, extending about 12 pixels in (0002) direction. The encircled regions were masked and used to define IFTs that are labeled RGB to follow the color coding used for creating the RGB image shown below in figure 3d.

We used the following method to determine that the lattice parameter differed from its average value in well-defined regions, and thus, to locate these strained regions. At each location of interest in the FT image, a Gaussian mask, 4 pixels in diameter, was placed manually, and the inverse FT (IFT) was performed on the masked region. Each IFT image obtained by this procedure displays the region in the real space image that has a lattice parameter corresponding to the centroid of the mask. For example, figure 2b shows masks at three locations, labeled B, G, and R, in the FT of the image in figure 2a. The associated IFT from each individual mask is shown in figures 3a-3c, respectively. An additive combination of the three IFT images, with the IFTs of the different masks labeled by color, is shown in figure 3d. Comparison of figure 3d with figure 2a shows that the region of largest lattice parameter (red-colored IFT) matches the location of the QW in the original real-space image. This result is consistent with the expectation of localized strain in and near the QW. The location of the centroid of each mask can be determined with sub-pixel precision. In the case of the masks at B, G, and R in figure 3b, the associated (0002) lattice spacings are 0.259 nm, 0.273 nm, and 0.288 nm, respectively.

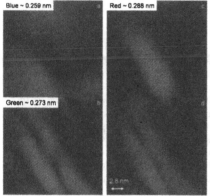

Figure 3. (a) IFT of the spot labeled B in Fig. 2b. This area represents the GaN component of the nanowire. (b) IFT of the spot labeled G in Fig. 2b. This area represents an area with a lattice spacing larger than GaN, and is thought to have a lower In concentration relative to the R spot. (c) IFT of the spot labeled R in figure 2b. This area has the largest lattice spacing. The respective values of the (0002) lattice spacings are indicated in figures 3a, 3b, and 3c. (d) Additive combination of the IFTs in figures 3a, 3b and 3c, with sources indicated by color (in the digital version of this paper). The variation of the (0002) lattice spacings in the QW is thought to be due to non-uniform In concentration.

High resolution TEM lattice structure, or dot, images were taken of GaN nanowires and InGaN QW by use of the C_s-corrected Titan at 300 kV. The image resolution was better than 0.08 nm in these images, as determined by analysis of the FTs. Figure 4a is from an unstrained region of the GaN nanowire and clearly shows the projected atomic columns of Ga in the wurtzite structure. Figure 4b is from the region containing the InGaN QW. The (0002) spots from the FT of both images are included for comparison and show clearly that the (0002) spot in the InGaN image is smeared out compared to the corresponding spot in the FT of the image of GaN only. Close inspection shows the presence of stacking faults and dislocations in the image of the QW, figure 4b, but they should not be the source of the observed smearing. The appearance of the FT and the associated lattice parameter results obtained from images taken by use of the two different TEMs were very similar.

Figure 4. (a) HRTEM lattice structure image (raw image) of an unstrained length of the GaN nanowire and, beneath, the (0002) spot in its FT. (b) HRTEM lattice structure image of a region around an InGaN QW and its (0002) FT spot. Note also the smearing of the (0002) spot in the FT from the QW relative to that from the GaN. The spot in the unstrained FT is about 5 pixels in diameter, while the spot in the QW is about 12 pixels in diameter, or more than two times larger. Analysis shows a maximum strain of 0.053 for this QW.

For each strain measurement, the lattice spacing along the nanowires axis associated with the central region of the QW was tabulated as d_{0002}. For clarity of description, we chose the lattice spacing of the GaN nanowire away from the quantum well as the unstrained reference value for all measurements. This choice is convenient because an accurate unstrained reference value of the lattice parameter throughout the InGaN QW is not known.

Thus, the strain in the <0002> direction is defined as

$$\epsilon_{0002} = \frac{(d_{0002}-d_{GaN})}{d_{GaN}} \quad , \tag{1}$$

where d_{GaN} is the (0002) lattice spacing of GaN, 0.25925 nm [11]. Repeated measurements on single images and measurements on different images of the same specimen indicated that the uncertainty of ϵ_{0002} is about ± 0.01.

RESULTS and DISCUSSION

The maximum strain in the <0002> direction in the InGaN QW shown in figure 2 is 0.11, as determined from the analysis described above and illustrated in figure 3. Results for six different nanowires, along with nanowire diameter, are listed in Table I. Similar analyses of smeared FT spots from images taken with the $g = \pm 1\bar{1}00$ three-beam conditions show that the strain in directions normal to <0002> is less than 0.005. This is consistent with the expected constraint imposed by the GaN lattice in directions normal to the <0002> growth direction.

Table I. Maximum measured values of axial strain in the InGaN QW (referred to unstrained GaN) and estimates of the strain based on three different physical assumptions.

	Maximum measured values of axial strain from FFT of QW						Vegard's law [13]	Pseudo-morphic [13]	Full volume change, axial strain only
Diameter, nm	60	87	88	102	109	110			
Strain	0.06	0.07	0.07	0.08	0.09	0.13	0.029	0.047	0.093

The present strain values for InGaN in GaN nanowires are larger than any found in the literature for other forms of $In_xGa_{1-x}N$ (QW)/GaN. Görgens et al. [13] show that the behavior calculated by a simple application of Vegard's law, that is, simple interpolation between the c-axis lattice parameters of unstrained GaN and InN, is quite different from that calculated for the pseudomorphic case, in which the elastic constraint provided by the GaN regions adjacent to the QW has a significant effect. The composition of a few QWs, obtained using EDS with a beam having a 10 nm radius, corresponds to 28±4 % InN in InGaN. This value was used to calculate the strain predicted for the two cases described by Görgens et al. [13], using a c-axis lattice spacing of 0.5718 nm [14]; the calculated strain values are listed in Table I.

If one assumes volumetric expansion of the QW according to Vegard's law for unconstrained crystals, and if at the same time its diameter is assumed to be constrained to that of the GaN, then the full volumetric expansion must occur along the nanowires axis. With these assumptions and our In concentration, the axial strain would be 0.093. This purely *ad hoc* model seems to provide an approximate description of the measured strain values.

An explanation for the tendency toward higher strain values for larger diameter nanowires could be pursued by considering the effect of the surface-to-volume ratio or possible non-uniformity in the In content. It has been reported [15] that planar InGaN QW with different thicknesses seemed to have different strains, but the experiments were much different from those of the present study. More measurements on different specimens and more detailed analysis are needed for understanding of the behavior of InGaN QW in GaN nanowires.

REFERENCES

[1]   Monemar, B. III-V nitrides - important future electronic materials, *Journal of Materials Science-Materials in Electronics* **10** (4), 227-254, 1999.

[2]   Jain, S. C., Willander, M., Narayan, J., Van Overstraeten, R. III-nitrides: Growth, characterization, and properties, *Journal of Applied Physics* **87** (3), 965-1006, 2000.

[3]   Orton, J. W., Foxon, C. T. Group III nitride semiconductors for short wavelength light-emitting devices, *Reports on Progress in Physics* **61** (1), 1-75, 1998.

[4]   Damilano, B., Grandjean, N., Dalmasso, S., Massies, J. Room-temperature blue-green emission from InGaN/GaN quantum dots made by strain-induced islanding growth, *Applied Physics Letters* **75** (24), 3751-3753, 1999.

[5]   Smeeton, T. M., Kappers, M. J., Barnard, J. S., Vickers, M. E., Humphreys, C. J. Electron-beam-induced strain within InGaN quantum wells: False indium "cluster" detection in the transmission electron microscope, *Applied Physics Letters* **83** (26), 5419-5421, 2003.

[6]   Bertness, K. A., Sanford, N. A., Barker, J. M., Schlager, J. B., Roshko, A., Davydov, A. V., Levin, I. Catalyst-free growth of GaN nanowires, *Journal of Electronic Materials* **35** (4), 576-580, 2006.

[7]   Bierwolf, R., Hohenstein, M., Phillipp, F., Brandt, O., Crook, G. E., Ploog, K. Direct Measurement of Local Lattice-Distortions in Strained Layer Structures by HREM, *Ultramicroscopy* **49** (1-4), 273-285, 1993.

[8]   Jouneau, P. H., Tardot, A., Feuillet, G., Mariette, H., Cibert, J. HREM Strain Measurement of Ultra-Thin ZnTe and MnTe Layers Grown in CdTe, *Microscopy of Semiconducting Materials* (134), 329-332, 1993.

[9]   Hytch, M. J., Snoeck, E., Kilaas, R. Quantitative measurement of displacement and strain fields from HREM micrographs, *Ultramicroscopy* **74** (3), 131-146, 1998.

[10] Robertson, M. D., Corbett, J. M., Webb, J. B., Jagger, J., Currie, J. E. Elastic strain determination in semiconductor epitaxial layers by HREM, *Micron* **26** (6), 521-537, 1995.

[11] Robins, L. H., Bertness, K. A., Barker, J. M., Sanford, N. A., Schlager, J. B. Optical and structural study of GaN nanowires grown by catalyst-free molecular beam epitaxy. I. Near-band-edge luminescence and strain effects, *Journal of Applied Physics* **101** (11), 2007.

[12] Commercial equipment, instruments, or materials are identified only in order to adequately specify certain experimental procedures. In no case does such identification imply recommendation or endorsement by the National Institute of Standards and Technology, nor does it imply that the products identified are necessarily the best available for the purpose.

[13] Görgens, L., Ambacher, O., Stutzmann, M., Miskys, C., Scholz, F., Off, J. Characterization of InGaN thin films using high-resolution x-ray diffraction, *Applied Physics Letters* **76** (5), 577-579, 2000.

[14] Wright, A. F.; Nelson, J. S. Consistent Structural-Properties for Aln, Gan, and Inn, *Physical Review B* **51** (12), 7866-7869, 1995.

[15] Lin, Y. S. Study of various strain energy distribution in InGaN/GaN multiple quantum wells, *Journal of Materials Science* **41** (10), 2953-2958, 2006.

Mater. Res. Soc. Symp. Proc. Vol. 1184 © 2009 Materials Research Society   1184-HH01-09

Chemical Nano-Tomography of Self-Assembled Ge-Si:Si(001) Islands From Quantitative High Resolution Transmission Electron Microscopy

Luciano A. Montoro[1], Marina S. Leite[1], Daniel Biggemann[1], Fellipe G. Peternella[1], K. Joost Batenburg[3], Gilberto Medeiros-Ribeiro[1,2] and Antonio J. Ramirez[1]

[1]Brazilian Synchrotron Light Laboratory, P.O. Box 6192, Campinas SP 13084-971, Brazil
[2]Hewlett-Packard Laboratories, P.O. Box 10350, Palo Alto, California 94303-0867, USA
[3]University of Antwerp, Universiteitsplein 1, B-2610, Wilrijk, Belgium

ABSTRACT

The knowledge of composition and strain with high spatial resolution is highly important for the understanding of the chemical and electronic properties of alloyed nanostructures. Several applications require a precise knowledge of both composition and strain, which can only be extracted by self-consistent methodologies. Here, we demonstrate the use of a quantitative high resolution transmission electron microscopy (QHRTEM) technique to obtain two-dimensional (2D) projected chemical maps of epitaxially grown Ge-Si:Si(001) islands, with high spatial resolution, at different crystallographic orientations. By a combination of these data with an iterative simulation, it was possible infer the three-dimensional (3D) chemical arrangement on the strained Ge-Si:Si(001) islands, showing a four-fold chemical distribution which follows the nanocrystal shape/symmetry. This methodology can be applied for a large variety of strained crystalline systems, such as nanowires, epitaxial islands, quantum dots and wells, and partially relaxed heterostructures.

INTRODUCTION

The growth of 3D semiconductor islands in heteroepitaxial systems and the study of their physical properties have been a very active area of research, principally because of their optical and electronic properties. An accurate control of the shape, size, composition and elastic strain of the islands is a crucial step for the application of these nanostructures in the quantum devices. Fluctuations on the composition inside of these nanostructures are a key factor that determine their structural properties and, as a consequence, the associated physical properties [1]. The Ge-Si islands grown on Si(001) have emerged as prototypical model system for the study of self-assembled semiconductors nanocrystals. This assumption stems from the fact that only two elements are involved, which are completely miscible in one another, and for exploring the conventional Si technology. The properties of Si-Ge:Si(001) have been studied extensively using a number of techniques, including atomic force microscopy combined with selective chemical etching [2], X-ray absorption fine structure [3], transmission electron microscopy (TEM) [4,5] and grazing incidence anomalous X-ray diffraction [6,7]. All support the existence of vertical and radial composition variations, with most of the Si at the base of the island and the Ge concentration increasing monotonically from the base to the top.

Scanning Transmission Electron Microscopy (STEM) has also been a useful tool, providing individual islands chemical compositional profiles from Energy Dispersive X-ray

Spectroscopy (XEDS), and Electron Energy Loss Spectroscopy (EELS) [5,8]. However, the results available in the literature obtained using these techniques provide limited information about the composition of the whole island.

On the other hand, more detailed information can be provided by advanced QHRTEM techniques. In the last decade several methods have been proposed and applied to different systems to obtain crystallographic properties, strain, stress and chemical composition from high-resolution TEM images. A comprehensive review on the different methods and program packages of QHRTEM is given by *Kret et al* [9]. A useful method for strain and chemical composition analysis is based on lattice parameter measurement. For some binary ($A_{1-x}B_x$) and pseudo-binary ($A_{1-x}B_xC$) bulk materials the Vegard's law assumes a linear relationship between the lattice parameter and the chemical composition. This technique has been successfully applied to the investigation of semiconductors quantum well layers, incoherently III-V strained islands and quantum dots. [9,10]

Here we combine Geometrical Phase Analysis (GPA) [11] and focal series aberration corrected images obtained by High Resolution Transmission Electron Microscopy (HRTEM) at two orientations on 40 ± 5nm diameter Ge_xSi_{1-x}:Si(001) islands, allowing the investigation of their 3D chemical composition in a quantitative fashion.

GPA is a method for measuring and mapping structural displacement fields on HRTEM images using a reference lattice. This method has been effectively employed by *Hÿtch et al.* [12,13] to study strain fields with resolutions down to 0.001 nm in semiconductors and metals. The focal series reconstruction (FSR) is a numerical technique for the restoration of the phase and the amplitude of the exit-plane wavefunction from a series of images obtained with different defocus values [14]. By this means, objective lens aberrations-free images with improved resolution that nearly achieve the instrumental information limit can be obtained. This exit-plane wave reconstruction procedure applied together with GPA improves the accuracy of the quantitative determination of the displacement fields, reducing the effect of the aberrations introduced by the microscope imaging system [15,16]. In addition, the minor distortions caused in the images by the TEM projector lenses, which could compromise the quantification of the displacement fields, have been corrected using a GPA based procedure [17].

This methodology has been applied to obtain the Ge-Si islands distortion maps and from them, their respective lattice parameter components (parallel, $a_{//}$ and perpendicular, $a_{\perp}$). Composition and strain components ($\varepsilon_{//}$ and $\varepsilon_{\perp}$) maps were determined in a self-consistent manner by relating the lattice parameters ($a_{//}$ and $a_{\perp}$) by anisotropic elastic theory using the composition-dependent stiffness coefficients, assuming biaxial-stress and Vegard´s law.

THEORY AND EXPERIMENTAL SECTION

Theoretical Basis:

The geometrical phase analysis (GPA) technique can be used to the determination of the lattice displacement field $u(r)$ at specific directions. This displacement from their reference periodic position is determined from the phase term of a set of lattice fringes analyzed as a Fourier series. In this case, the reference set corresponds to an undistorted pure silicon region of the substrate far from the domes. The lattice distortion or strain tensors can be calculated by the gradient of the displacement field as follows:

$$e_{ij} = \frac{1}{2}\left(\frac{\partial u_i}{\partial x_j} + \frac{\partial u_j}{\partial x_i}\right) \qquad (1)$$

The GPA method provides the lattice distortion maps ($e_{//}$, parallel to the substrate and $e_\perp$, normal to the substrate) from a Si reference lattice, and these distortion components can be defined in terms of the lattice parameters by:

$$e_{ij} = \frac{a_{ij}}{a_{ref}} - 1 \qquad (2)$$

where a_{ij} and a_{ref} are the distorted lattice parameter components ($a_{//}$ and $a_\perp$) and the silicon reference lattice parameter ($a_{Si} = 0.5430$ nm), respectively. From the $a_{//}$ and $a_\perp$ maps, the chemical composition of the Ge_xSi_{1-x} (where $0 \geq x \geq 1$) islands can be calculated through the Vegard's law,

$$a_{(x)} = a_{Si}(1-x) + a_{Ge}x \qquad (3)$$

and the strain tensor components, calculated from the equation:

$$\varepsilon_{ij} = \frac{a_{ij}}{a_{(x)}} - 1 \qquad (4)$$

Thus, considering the islands a thin epitaxial system with a free surface, the perpendicular stress can be disregarded (plane-stress) resulting in a useful relationship between perpendicular and parallel strains [18]:

$$\varepsilon_\perp = \frac{-2C_{12}}{C_{11}}\varepsilon_{//} \qquad (5)$$

Combining equations (4) and (5), and applying Vergard's law to correct the composition dependent parameters ($a_{(x)}$ and $C_{ij(x)}$) we find the final quadratic expression for the composition:

$$\frac{a_\perp - [a_{Ge}x + a_{Si}(1-x)]}{a_{//} - [a_{Ge}x + a_{Si}(1-x)]} + 2\frac{[C_{12}^{Ge}x + C_{12}^{Si}(1-x)]}{[C_{11}^{Ge}x + C_{11}^{Si}(1-x)]} = 0 \qquad (6)$$

The Ge-Si alloy presents a high elastic anisotropy. Thus, in order to calculate the stiffness coefficients tensors for an arbitrary crystal orientation one must perform a transformation using a rotation matrix [19].

Therefore, from the distortion maps provided by GPA method, the stiffness coefficients calculated for each analyzed direction, and considering the values of $a_{Si} = 0.5430$ nm, $a_{Ge} = 0.5657$ nm, the Ge content x was calculated finding the zero of equation (6). Thus, the chemical composition maps were obtained using a self-consistent methodology assuming that the anisotropic elastic theory could be applied to such nanostructures. The value of composition obtained for each atomic column of the HRTEM image represents the mean chemical composition along the corresponding direction with 1.6 nm spatial resolution, as imposed by the Lorentzian mask used in the GPA.

Materials and Methods:

Ge-Si islands were grown on Si (001)-oriented substrate by a Chemical Vapor Deposition (CVD) technique at a total pressure of 1.3 kPa (10 torr) in a lamp-heated, single wafer reactor. An in-situ surface preparation was conducted by heat treatment at about 1150 °C in a H_2 ambient and deposition of a Si buffer layer at 1080 °C. The Ge was deposited using GeH_4 with a partial pressure of 0.065 Pa in a H_2 carrier gas for 120 seconds at 605 °C, resulting in a final thickness of ~ 12 eq-ML (1 equivalent-monolayer = $6.27 \cdot 10^{14}$ Ge atoms/cm^2) [20]. AFM statistical analysis pointed out uniform dome-shaped islands with a narrow size distribution.

TEM cross-section specimens oriented at [110] and [100] zone axis were prepared using manual and dimpler polishing followed by liquid nitrogen cooled Ar$^+$ ion-beam thinning with energies of 3.5 keV and 2 keV at incidence angles gradually decreasing from 6° to 2°. A JEM-3010 URP TEM with a LaB$_6$ electron gun and spherical aberration coefficient of 0.7 mm was used at an accelerating voltage of 300 kV.

The distortions introduced into HRTEM images by the projector lens system were removed using the procedure proposed by Hüe et al. [17]. Spherical aberration corrected HRTEM images were obtained using exit-plane wave function restoration from focal series technique. The procedure used for the restoration was the Wiener Filter method [17], implemented as a routine (FTSR from *HREM Research Inc.*) written for the software package Digital Micrograph™. Sets of images for focal series reconstruction were recorded on a 1024×1024 thermoelectrically cooled CCD camera with a focal step of 5 nm between exposures. An adequate magnification was chosen to obtain a sampling rate of ~0.030 nm per pixel; i.e. an image discretization well below the Nyquist limit of 0.077 nm, excluding aliasing phenomena. The image processing by the geometric phase analysis [11] was carried out using a routine (FTSR from *HREM Research Inc.*) implemented on Digital Micrograph™, and was applied to the amplitude image obtained from the reconstructed exit wave. For the GPA processing a Lorentzian mask was used with a half-width of $1 / 4^*d_{111}$ in reciprocal space, resulting in 2D chemical and strain maps with 1.6 nm spatial resolution. The precision in the chemical quantification was estimated for each map based on the standard deviation of the measured composition in a Ge-free region of the silicon substrate. Important aspects of the images considered for GPA are the absence of contrast reversals in the image through the sample thickness and the absence of significant delocalization contrast [15]. The thickness variation in ~40 nm diameter Ge-Si islands did not cause contrast reversals at the used defocus.

RESULTS AND DISCUSSION

Figures 1A and 1B show the 2D chemical maps of two representative islands obtained self-consistently from the high resolution images of the reconstructed exit-plane wavefunctions amplitude along the [110] and [100] directions, respectively. These chemical maps are side-view projections of the islands, representing the values averaged through the atomic columns. The analysis of the chemical maps obtained from several islands in different TEM specimens revealed a good reproducibility. In addition a discerned examination of the HRTEM images and chemical maps was performed to avoid the analysis of severely sectioned or sliced islands originated from specimen preparation. These analyses were performed in at least 3 different TEM specimens for each direction, totalizing 6 to 7 islands with similar profiles. However, small variations among the chemical maps are expected due to specimen preparation issues, defects on the nanocrystals faceting, and the intrinsic QHRTEM method errors sources.

The chemical map at the [110] direction (Fig. 1A) shows a non-monotonic behavior along the [001] direction for the Ge distribution (%$_{atom}$). This projection shows a Ge-enrichment at the edges (65 ± 8 %Ge), at the core near to the base (70 ± 8 %Ge), and at the top (90 ± 8 %Ge) of the island, with a region between them with 45 ± 8 %Ge. Contrasting with Fig. 1A, Fig. 1B ([100] direction) shows a Si-enrichment at the edges (28 ± 8 %Ge) and a nearly uniform composition at the central area (58 ± 8 %Ge), while a high Ge concentration was also found on the top of the islands (90 ± 8 %Ge). These features show that the [110] and [100] projections are significantly different, and clearly suggest that the Ge arrangement is non-cylindrically symmetrical.

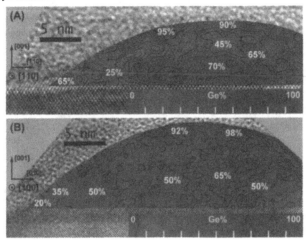

Figure 1. Two-dimensional chemical maps of representative Ge-Si:Si(001) dome-shaped islands, where (A) and (B) corresponds to the chemical maps obtained along the [110] and [100] crystallographic directions, respectively. The color code indicates the Ge content. The maps are superposed to the original HRTEM images, showing the coherently-strained Ge-Si islands on the Si(001) substrate. The indicated values present a deviation of ±8%, which were estimated from the standard deviation of the measured compositions at the Si lattice reference.

The 2D chemical maps correspond to the projected average composition at each point, allowing to unveil 3D features. Thus, by using selected projections it is possible infer the 3D chemical distribution in a self-consistent manner. Due to the structural symmetry of the islands, two distinct projections ([110] and [100]) from statistically representative islands were sufficient for performing the 3D reconstruction.

Figure 2 shows the 4-fold symmetric model for the chemical distribution, derived from an interpretation of the 2D projected chemical maps and the crystal symmetry. Ge-Si islands are multi-faceted structures bounded by the {113}, {105} and {15 3 23} planes [21], as schematically shown on Fig. 2. The chemical distribution inside the island can be related to the nanocrystal shape due to this feature determines the strain relaxation. The islands chemical distribution can be associated to the nanocrystal shape due to these features are interrelated to thermodynamic and stress relaxation effects. This relationship can be attributed to interdiffusion processes induced by a complex balance among the elastic energy minimization, kinetic, and/or thermodynamic effects [22].

The [110] projected chemical map (Fig. 1A) suggests a Ge enrichment at the bottom of the {113} facets and at the island top. Moreover, the [100] chemical map (Fig. 1B) indicates a Si-rich region at the bottom of the {15 3 23} facets and a Ge enrichment at the top. This 3D island model was constructed for validate the proposed four-fold symmetry and determine an approximated geometry for the chemical distribution. An algorithm was used to implement this model, which was formed by intersections of different functions, resulting in the structure shown on Fig. 2, where constant composition volumes are highlighted in red (Ge-rich) and green (Si-rich). The 4-fold 3D model of the island is formed by Ge-enriched regions located by the {113} facets (88 %Ge), at the dome top (90 %Ge), and at the island core (75 %Ge); Si-rich regions by the {15 3 23} facets (25 %Ge) and at the base (25 %Ge), with the volume among the above parts filled by 50 %Ge.

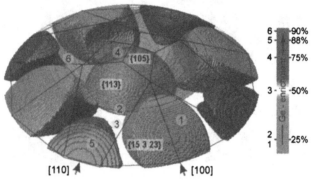

Figure 2. 3D chemical distribution of the Ge-Si:Si(001) dome-shaped islands. The line boundaries show the faceted dome, where the facets families (braces) and the directions (brackets) are indicated. The colors differentiate Ge-rich (reddish) and Si-rich (green) regions. The numbers attributed to each region group refers to specific Ge content (%$_{atom}$), as indicated by the scale bar.

From this model the 2D projected averaged maps of composition were calculated for the [110] and [100] projections. By comparison of these 2D projected views with the respective chemical maps, the 3D model was iteratively refined optimizing shape and composition, resulting in the representation of the chemical distribution shown on Fig. 2.

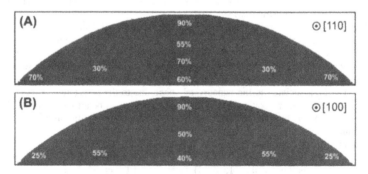

Figure 3. 2D projected averaged composition maps of the 3D model island. (A) and (B), show the projected chemical maps obtained for the [110] and [100] directions, respectively. The values indicate the averaged composition at each region.

Figure 3 shows the projected maps obtained from the refined model along the [110] and [100] directions. A qualitative visual comparison of these maps with the experimental chemical maps from Fig. 1 shows a good agreement. In addition, a quantitative evaluation of the refined model from the values of composition indicated on Fig. 3, shown that the experimental data are in good accordance with the 3D model for both directions. The vertical Ge variation at the [110] direction, Fig. 3A, shows the previously mentioned non-monotonic profile, which results from the projected 3D chemical distribution. These results highlight the importance of quantitative measurements of 3D composition of nanostructures, in which complex chemical profiles may be present.

In summary, we report a methodology that uses HRTEM images to obtain 2D chemical maps in a self-consistent fashion, and a reconstruction of the 3D composition arrangement of Ge-Si:Si(001) islands. This methodology stands out as a high spatial resolution tool for obtaining 3D quantitative chemical information, which can be applied to numerous alloyed nanostructured strained crystalline systems.

ACKNOWLEDGMENTS

The authors acknowledge LNLS, FAPESP (2007/05165-7) CNPq (382850/2007 - 6; 152445/2007-2), and HP Brazil for the financial support. GMR acknowledges Dr. T. I. Kamins for kindly supplying the sample utilized in this work.

REFERENCES

1. D. Bimberg, et.al, *Quantum Dot Heterostructures* (John Wiley & Sons: Chichester, 1999).
2. A. Rastelli, M. Stoffel, A. Malachias, T. Merdzhanova, G. Katsaros, K. Kern, T. H. Metzger, O. G. Schmidt, *Nano Lett.* **8**, 1404 (2008).
3. F. Boscherini, G. Capellini, L. Di Gaspare, F. Rosei, et.al, *Appl. Phys. Lett.* **76**, 682 (2000).
4. E. P. McDaniel, P. A. Crozier, J. Drucker, D. J. Smith, *Appl. Phys. Lett.* **87**, 223101 (2005).
5. S. A. Chaparro, J. Drucker, Y. Zhang, D. Chandrasekhar, M. R. McCartney, D. J. Smith, *Phys. Rev. Lett.* **83**, 1199 (1999).
6. A. Malachias, S. Kycia, G. Medeiros-Ribeiro, R. Magalhães-Paniago, T. I. Kamins, R. S. Williams, *Phys. Rev. Lett.* **91**, 176101 (2003).
7. T. U. Schülli, J. Stangl, Z. Zhong, R. T. Lechner, M. Sztucki, T. H. Metzger, G. Bauer, *Phys. Rev. Lett.* **90**, 066105 (2003).
8. M. Schade, F. Heyroth, F. Syrowatka, H. S. Leipner, T. Boeck, M. Hanke, *Appl. Phys. Lett.* **90**, 263101 (2007).
9. S. Kret, P. Ruterana, A. Rosenauer, D. Gerthsen, *Phys. Stat. Sol. B* **227**, 247 (2001).
10. A. Rosenauer, *Transmission Electron Microscopy of Semiconductor Nanostructures: Analysis of Composition and Strain State* (Springer-Verlag: Berlin, 2003).
11. M. J. Hÿtch, E. Snoeck, R. Kilaas, *Ultramicroscopy* **74**, 131 (1998).
12. M. J. Hÿtch, J.-M. Putaux, J.-M. Penisson, *Nature* **423**, 270 (2003).
13. C. L. Johnson, E. Snoeck, M. Ezcurdia, B. Rodríguez-González, I. Pastoriza-Santos, L. M. Liz-Marzán, M. J. Hÿtch, *Nature Materials* **7**, 120 (2008).
14. R. R. Meyer, A. I. Kirkland, W. O. Saxton, *Ultramicroscopy* **92**, 89 (2002).

15. J. Chung, L. Rabenberg, *Ultramicroscopy* 108, 1595 (2008).
16. M. J. Hÿtch, T. Plamann, *Ultramicroscopy* 87, 199 (2001).
17. F. Hüe, C. L. Johnson, S. Lartigue-Korinek, G. Wang, P. R. Buseck, M. J. Hÿtch, *J. Electron Microscopy* 54, 181 (2005).
18. J. Y. Tsao, *Materials Fundamentals of Molecular Beam Epitaxy* (Academic Press, 1993)
19. J. J. Wortman, R. A. Evans, *J. Appl. Phys.* 36, 153 (1965).
20. T. I. Kamins, E. C. Carr, R. S. Williams, S. J. Rosner, *J. Appl. Phys.* 81, 211 (1997).
21. F. M. Ross, R. M. Tromp, M. C. Reuter, *Science* 286, 1931 (1999)
22. U. Denker, M. Stoffel, O.G. Schmidt, *Phys. Rev. Lett.* 90, 196102 (2003).

Mater. Res. Soc. Symp. Proc. Vol. 1184 © 2009 Materials Research Society   1184-HH02-03

Hybrid Tomography of Nanostructures in the Electron Microscope

Z. Saghi, T. Gnanavel, X. J. Xu, and G. Möbus
Department of Engineering Materials, University of Sheffield, Sheffield, S1 3JD, UK

ABSTRACT

A variety of tomographic experiments and modes for electron tomography of nanostructures are introduced, derived from the general concepts of quantitative computed tomography, binarised geometric tomography, including shape-from-silhouette, and spectroscopic chemical mapping. Our emphasis is on working out concepts of combining at least two of these tomography modes in order to share their respective advantages and improve the overall reconstruction quality. In this work, the following three hybrid modes are presented: (i) ADF-STEM tomography and EDX tomography into high-resolution 3D chemical mapping, (ii) geometric tomography and lattice-resolved backprojection into HREM-tomography for convex bodies, and (iii) geometric tomography and e-beam nanosculpting into "tomographic nanofabrication".

INTRODUCTION

Electron tomography in the transmission electron microscope (TEM) [1] is the oldest form of "nanotomography" reaching nanometer-scale resolution in 3D from its beginnings in 1968. After three decades of tomography based on bright field conventional TEM imaging, the last ten years have seen a large number of new signals being introduced for electron tomography.

We classify all the old and new modes into major groups (Fig 1), separating the cases of:

(i) Quantitative density mapping (equivalent to x-ray computed tomography, CT), where the signal is proportional to a materials property, e.g. density, integrated along the projected thickness. The absorption index is the most common example and images can be easily inverted or linearised by applying a logarithm function (see Fig 1b, top line).

(ii) Z-contrast mapping is a special case of (i) where the signal is proportional to atomic number times density. However, different from case (iii), multiple elements are integrated/mixed rather than isolated in compound materials. Examples of this category are high angle annular dark field scanning TEM (HAADF-STEM) [2] or weak phase object phase contrast imaging of thin specimens (see Fig 1b, bottom line).

(iii) Spectroscopic imaging, where the signal is the integrated density of one element only, even if part of a compound crystal. This has been realized by energy filtered TEM (EFTEM) [3], electron energy loss spectroscopy (EELS) spectrum imaging, and energy dispersive X-ray (EDX) mapping [4]. Fig 1d,e show the two projections of a core-shell particle with the spectrometer tuned to each element.

(iv) Silhouette or shadow mapping is a case of "Geometric Tomography" [5] and operates on binarised signals (0 or 1) derived from any of the above modes (see fig 1c). The backprojection of binarised projection images leads to an approximate solution for the 3D structure with the following features:

- Any internal density fluctuations are ignored. Only a surface is reconstructed. The method therefore is particularly powerful for constant density nanoparticles.

- The true surface is not always recovered, but approximated by the "Convex Hull", a body which can be derived for a random 3D surface by replacing all surface elements by tangential planes which do not intersect the body at any other position. In practice, this means replacing "dents, dimples or cracks" or any other concave surface morphology by smooth planes leaving a convex body behind. If the original surface is already convex, then the convex hull approaches the true surface the finer the tilt increment. As an example, we show in Fig 1f four binary projections of a square and the volume intersection (in blue) of the backprojected shadows (in green) which approximated the original test object shape.

- For axial tomography in particular, where the volume is reconstructed slice by slice, the above convexity criterion can be restricted to individual cross-sections, which allows for some concave objects to be perfectly retrieved as long as all cross-sections perpendicular to the rotation axis are convex.

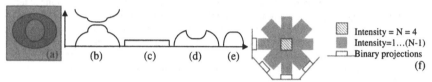

Figure 1. A model core-shell object of different chemistry (a) and five tomography modes (b-e) suitable for hybrid electron tomography (see text). (f) Scheme of geometric tomography (silhouette backprojection) for an axially convex object: hashed (blue on-line) is the intersection of the 4 backprojected shadows.

HYBRID TOMOGRAPHY MODES

It is obvious that the modes of Fig 1 provide a rich variety of advantages and disadvantages, and that the recording of two different tomograms, to be combined at the end (by addition, multiplication or logical operators) into one single tomogram, can lead to superior results. We discuss 3 combinations of pairs of modes into hybrid schemes, which look particularly promising. Historically, the idea of combined tomograms seems to stem from Tam [6], who proposed to combine two x-ray wavelengths to reconstruct objects with a thick core and thin external details, such that one wavelength optimizes the signal-to-noise ratio for the inner microstructure of the core, and another wavelength optimizes sensitivity and visibility for the thin marginal regions.

Combination of structural and spectroscopic tomography modes

Resolution of chemical maps by EFTEM or EDX-mapping is inherently limited to worse than 1nm due to delocalization of scattering, bad signal-to-noise, and long acquisition/scanning artefacts. In our earlier work, we realized that since the spatial definition of an EDX scan is performed on ADF-STEM images anyway, it would be advantageous to combine an EDX chemical mapping tomogram with a reconstructed ADF-STEM tomogram revealing the external structure of the object [7]. Fig 2 summarizes our results using a tungsten tip covered with oxide and carbon layer as a core-shell test object, and tilted from -70° to +50° with 10° increment. Fig 2 a and b are EDX-linescan signals of tungsten, oxygen and carbon at -70° and +50° respectively. A red-green-blue map of the reconstructed chemical distribution is shown in fig 2 c, with the tungsten core in dark grey (blue on-line), the oxygen shell in light grey (red on-line) and the carbon layer in white (green on-line). Fig 2 d shows the geometric reconstruction of the same cross-section from binarised ADF-STEM projections. Interpreting this experiment in the wider context of hybrid tomography, we identify three main advantages of the combined tomogram:

- The ADF–STEM projections improve massively the precision of the lateral alignment of the stack of EDX-linescan projections due to its lower noise and higher contrast. In the case of 2D EDX mapping tomography [4], as the exposure time is about 1 sec for an ADF-STEM overview image, against 10min-1h for each EDX map, the tilt increment for the ADF-STEM mode could be chosen finer.

- The ADF-STEM images have a better structure resolution, and therefore can define the outer surface more accurately, while the EDX signal unambiguously identifies the chemical phases, which the ADF signal alone (even as HAADF Z-contrast) could not provide.

- The ADF-STEM imaging mode can be used to reduce star artefacts either via its finer tilt increment, or by reconstructing the binasired ADF-STEM projections into a convex hull multiplied to the chemical 3D tomogram. For convex objects this results in correct surface definition while leaving the interior chemical map unaffected.

Future extensions of this scheme could include combinations of EDX and EFTEM tomograms for elements which are either inaccessible by EDX or EELS due to atomic number.

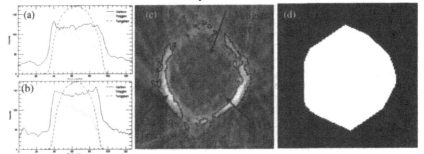

Figure 2. EDX line scan tilt series of C-K, O-K and W-Lα at -70°(a) and +50° (b). (c) RGB-map of the tomographic reconstruction of the chemical distribution of W-Lα (dark grey, blue on-line), O-K (light grey, red on-line) and C-K (white, green on-line), (d) shape-from-silhouette reconstruction of the cross-section in (c), from binarised ADF-STEM projections.

Combination of binary silhouette and lattice resolved tomography modes

A straight usage of high resolution TEM (HREM) lattice images for tomography is not possible, although simulations show that below ~3nm thickness recovery of atoms might be possible if resolution and first contrast extinction/ inversion thickness are high enough. However, HREM images can be the basis for extracting manually or automatically the silhouette /shadow as starting point for geometric tomography to get the convex hull (still providing atomic resolution surface definition) [8]. The backprojection of a selected subset of the HREM tilt series, including e.g. two or three zone axis oriented members, can then add valuable crystallography information to be superimposed to the convex hull. This idea had been formulated as part of the concept of lattice goniometry [9], and differs in purpose from weak-phase-object HREM tomography [10]. We selected a CeO_2 nanoparticle, in cubic fluorite structure, which was by chance aligned to the tomographic rotation axis, such that a {200} fringe system is visible throughout a tilt series from -60° to +70° with 10° increment. Fig 3 a-d show four members of the tilt series. The HREM images in fig 3a and fig 3b show zone-axis patterns of <110> type and near to <100> respectively, and were backprojected in full contrast, after noise filtering to select the major Bragg spots. All 14 images were binarised and reconstructed using the shape-from-silhouette algorithm, after segmentation to cut the particle of interest from its neighbour and to eliminate spurious patches of counts. Fig 3e shows the isosurface of the tomogram obtained from the binarised tilt series, while fig 3f shows the reconstruction by superposition of only two zone-axis projections.

The combination of the two is essential, as the HREM patterns alone do not give an alignment criterion, and the shared volume of the two backprojected HREM contrast patterns (both roughly cylindrical) do not reveal the particle shape, but merely a geometric artifact (common volume of two cylinders) depending on the tilt difference of the two zone axes. The geometric tomography however reveals the particle shape, as also shown in [11].

Figure 3 illustrates the concept of this hybrid method, without arriving at a reliable solution at the atomic scale. Conditions for fully successful operation would include:

- all HREM images must be at same focus value, and below an extinction/inversion thickness for the main fringe systems, and the lattice throughout the particle must be strain-free.
- all HREM images should be "Structure Images" in the sense of [12], which means that either all atoms, or at least the sublattice of heavy atoms, generate spots at the correct atomic positions.
- the surface contour images should allow alignment to the precision required to define the lateral positions of the atomic resolution images as well.
- all HREM images must be precisely angularly aligned not only to the rotation axis, but also with respect to the crystallographic zone axes. Small mistilts, as revealed by multislice simulations, can lead to lateral shifts of fringe patterns, or even pattern changes similar to focus and thickness drifts, e.g. due to weakening of non-linear contributions relative to linear ones (reduced dynamicity after O'Keefe [13]).

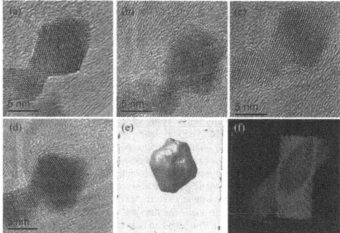

Figure 3. HREM projections of an isolated CeO2 nanoparticle, at nominal goniometer tilt angles of -50° (a), -10° (b), +10° (c) and +40° (d). (e) Shape-from-silhouette reconstruction from the full binarised tilt series.(f)Voxel projection view of the HREM reconstruction from the two zone-axis projections (a and b).

Tomographic nanofabrication

Although not a hybrid mode by multiplication of two tomograms, the focused electron beam patterning by hole drilling of a sample viewed in multiple directions using a high-tilt tomography holder fits into our scheme, as a second tomographic tilt series by imaging will reveal a 3D view of the freshly nanofabricated object [14]. In addition, if the 2nd step of reconstruction from images is performed by geometric tomography, we postulate a topological equivalence of the two processes of hole drilling and silhouette backprojection under the conditions:

- The electron beam is assumed to drill instant and straight holes all through the sample (no dose dependent partial hole thinning is allowed) thereby converting voxels of the original material into either deleted (voxel=0) or unchanged (voxel=1) material.

- The rotation axes of both processes are identical and the object belongs to a modified class of axially convex objects (as introduced in 1.): Rather than only requiring the external surface to be identical to its convex hull, we now allow for topologically multi-surfaced geometries. The internal surfaces of e.g. a cylindrical tube, lining the drilled hole, have to be axially convex, which refers to the flatness of each cross-section with respect to the rotation axis.

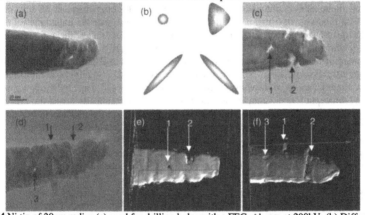

Figure 4 Ni tip of 20nm radius (a) used for drilling holes with a FEG e⁻ beam at 200kV. (b) Different shapes of the beams. The tip was tilted to -45° and drilled at the core of the tip and at three different positions at the edges (c). At +45°, the holes at positions 1 and 2 appeared as channels (d); an extra fine hole was drilled at position 3. A bright field -TEM tilt series was acquired from -45° to +45° with 5° increment. (e-f) Geometric tomographic reconstruction of the tip showing drilled holes and extrusions at positions 1 and 2.

As revealed by Fig 4, the rays of drilling realized by the focused electron beam play the same role than the rays of backprojection (inverting the parallel forward projection with a wide parallel electron beam), in separating filled from empty voxels. Upon superposition of multiple tilts, a voxel that has been set to 0 in a previous tilt angle, always remains 0, whether interpreted as drilling (you cant fill a hole by further drilling), or whether interpreted as silhouette backprojection (if one ray of light has passed the object without hitting voxels of 1, then all voxels along its line path remain 0).

Our experimental illustration of this hybrid tomographic nanofabrication scheme uses a nickel tip (fig 4a) which is cut (using ultra-astigmatic line focus) and drilled (using focused beams) (fig 4b) to perforate some cross-sections and to sharpen the very tip. Fig 4c-d show drilled holes at 3 different positions, while fig 4 e-f are the isosurface views of the reconstructed tip (using IMOD [15]) showing the overall shape and extrusions. Promising applications could be imagined for magnetic force microscopy tip sharpening, or to provide anchor points for carbon nanotube pickup or similar.

Experimentally, the above scenario will become modified to some degree: (1) materials redeposition after ablation will fill some voxels supposed to be empty, even outside the original silhouette before drilling starts; (2) the Gaussian beam shape leads to some partial-depth drilling near the walls of holes due to the tails of the beam. For one beam position therefore voxels of 1 and 0 could be mixed along its line path; (3) amorphisation, recrystallisation, and other materials density fluctuations would lead to deviations from the assumption of a constant density single-phase material.

CONCLUSIONS

We showed in this work how tomograms from different imaging modes can be combined to improve the resolution of the reconstruction and/or provide complementary 3D information. ADF-STEM binary reconstruction of a thick tip was used to estimate the boundaries of the cross-section reconstructed by EDX linescan tomography and therefore reduce the artefacts due to a large tilt increment. This approach can be extended enhancing EDX-mapping resolution via a finer tilt increment selected for ADF-STEM. The combination of HREM tomography with binary reconstruction was also explored as a mean to extract 3D crystallographic information from few zone-axis projections while estimating the overall crystal shape by binary tomography. A proof of concept of the technique was illustrated on a CeO_2 nanoparticle and the experimental challenges related to the technique were pointed out. Finally, tomographic nanofabrication was introduced as a technique to sculpture nanomaterials from different tilt angles using a tomography holder, and subsequently reconstructing the freshly machined object using the same instrumentation. The proposed combination of tomograms takes a full advantage of the capabilities of the instrument and therefore optimizes the extracted 3D information.

REFERENCES

1. Frank J (Ed.) 2007 *Electron Tomography: Methods for Three-Dimensional Visualization of Structures in the Cell, 2nd ed.* (New York:Plenum Press).
2. Midgley P A and Weyland M 2003 *Ultramicroscopy* **96** 413.
3. Möbus G and Inkson B J 2001 *Appl. Phys. Lett.* **79** 1369.
4. Möbus G and Inkson B J 2003 *Ultramicroscopy* **96** 433.
5. Gardner R J 1995 *Geometric Tomography* (New York:Plenum Press).
6. Tam KC 1987, *J nondestructive evaluation*, **6** 189.
7. Saghi Z, Xu X J and Möbus G 2007 *Appl. Phys. Lett* **91** , 251906.
8. Xu X J, Saghi Z, and Möbus G 2007 *MRS Symp.Proc.* **1026E** 1026-C08-04.
9. Qin W and Fraundorf P 2003 *Ultramicroscopy* **94** 245.
10. Möbus G, Doole R and Inkson BJ 2003 *MRS Symp.Proc.* **738**, G1.2, 15.
11. Xu X, Saghi Z, Gay R and Möbus G 2007 *Nanotechnology* **18** No 22 225501.
12. Spence JCH 1988 *Experimental High-Resolution Electron Microscopy* (New York: Oxford University Press).
13. O'Keefe M A and Radmilovic V 1993 *Proceedings - Annual Meeting, Microscopy Society of America* 980.
14. Saghi Z, Gnanavel T, Peng Y, Inkson B J, Cullis A G, Gibbs M R and Möbus G 2008 *Appl. Phys. Lett.* **93** 153102.
15. Kremer J R, Mastronarde D N and McIntosh J R 1996 *J. Struct. Biol.* **116** 71.

Mater. Res. Soc. Symp. Proc. Vol. 1184 © 2009 Materials Research Society 1184-HH02-08

3D Quantitative Characterization of Local Structure and Properties of Contact Materials

A. Velichko, M. Engstler, C. Selzner and F. Mücklich
Institute of Functional Materials, Department of Materials Science and Engineering, Saarland University, Saarbruecken, Germany

ABSTRACT

The characterization of spatial distribution of different phases in materials provides understanding of structural influence on the properties and allows making physically well-grounded correlations. The FIB/SEM nanotomography opens new possibilities for the target microstructure characterization on the scales from 10 nm to 100 μm. It is based on the automatic serial sectioning by the focused ion beam (FIB). Scanning electron microscope (SEM) in high resolution mode can be used for the imaging of nanostructured materials. Afterwards a detailed three dimensional (3D) image analysis enables the comprehensive quantitative evaluation of local microstructure. The possibilities of these techniques will be presented on the example of silver-composite contact materials which were analyzed using FIB nanotomography before and after exposure to plasma discharge. Significant changes in the spatial distribution of the oxide particles within the switched zone induce among other effects the changes in the local electric and thermal properties. These cause eventually the failure of the contact material. Advanced methods of image analysis allow characterization of inhomogeneous distribution of oxide particles in silver contact materials. Quantitative parameters characterizing the agglomeration of oxide inclusions and accumulation of pores can be derived from the results of distance transformations and morphological operations. The additional consideration of the connectivity allows the quantification of homogeneous and inhomogeneous states with high sensitivity and confidence level. Local thermal and electrical properties were estimated using simulation software on the real tomographic data. The combination of FIB microstructure tomography with modern 3D analysis and simulation techniques provides new prospects for targeted characterization and thus understanding of the microstructure formation and local effect associated with e.g. electro-erosion phenomena [1], [2]. First correlations between 3D microstructure parameters and resulting properties will be discussed.

INTRODUCTION

The properties of many engineering materials are strongly influenced by their microstructures. In the case of simple arrangements of microstructural constituents, like statistically distributed, spherical inclusions or uniaxial fibers, a 2D section is sufficient for the characterization of the materials microstructure. This approach fails, as soon as the phases are not uniformly distributed and of complex or non-convex shape. Although conventional 2D image analysis provides statistical results about the microstructure, the complete information about the 3D arrangement is not accessible. Unambiguous characterization of complex microstructures can only be done with the help of 3D analysis.

EXPERIMENT

In this work, erosion behavior of silver-based contact materials has been studied. Electro erosion craters have been generated by disconnecting electrodes under direct current. 3D analyses of the microstructures have been done by FIB-nanotomography, using a dual beam workstation (FEI Strata DB 235). In FIB-nanotomography the region of interest (ROI) is consecutively sliced at equal distances with the focused ion beam (FIB) and each cross section is imaged with the electron beam (SEM). Imaging can also include chemical contrast (energy dispersive x-ray spectroscopy, EDS) [3] and crystallographic orientation (electron backscatter diffraction, EBSD) [4]. The set of 2D images is reconstructed using Amira 4.0 software. In this work, two ROIs of approximately $10\times10\times10$ μm^3 of a reference sample and an electro-eroded (switched) sample have been analyzed. Voxel size is $18\times23\times40$ nm^3.

The reconstructed microstructures of these two samples of an Ag/SnO2 contact material have been analyzed using 3D image analysis software MAVI (Modular Algorithms for Volume Images). The algorithms for calculating the characteristic parameters of the microstructures are the same as described in [5], [6]. MAVI software does not only calculate parameters for the whole microstructure [7] but also allows image processing e.g. morphological operations and distance transformations. Using these methods, the changes in microstructure can be described quantitatively.

Electrical conductivity, one of the most important properties of contact materials, has been estimated in 3D using GeoDict simulation software [8], which is able to deal with real tomographic data.

DISCUSSION

FIB-nanotomography and reconstruction

The first reconstruction shows a reference sample with a microstructure in an initial state (figure 1). The second reconstruction was done in the switched area, 70 μm below the surface, where the segregation of oxides and formation of pores occurs (figure 2).

Figure 1. Sample in initial state. FIB cross section in secondary electron contrast (left), reconstructed oxides (middle) and reconstructed pores (right)

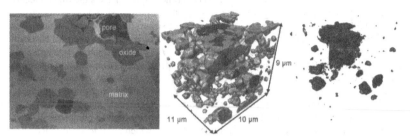

Figure 2. Eroded sample. FIB cross section in secondary electron contrast (left), reconstructed oxides and pores (middle) and reconstructed pores (right)

In the initial material oxides and pores are homogeneously distributed. Small pores are often located between two SnO_2 particles or two grains of the matrix. Formation of pores at the interface between matrix and oxides results from the production process of these materials. In the electro eroded sample, oxides are concentrated around larger pores. These pores have appeared due to the electro erosion process. Small pores are not surrounded by oxides.

3D image analysis

Four field parameters build the basis for the quantitative image analysis: volume fraction (V_V), surface area density (S_V), integral of mean curvature density (M_V) and integral of total curvature density (K_V). With its help, the microstructure can be described completely. Whereas V_V describes the composition of the material, S_V is predominantly influenced by kinetic aspects of the materials fabrication. M_V and K_V describe the geometric configuration (shape) of the different phases by summarizing all radii of curvature of the microstructure. V_V, S_V and M_V can be determined from a 2D section using stereological equations, if the structure is isotropic. For anisotropic microstructures S_V and M_V can only be determined from tomographic data. 3D data is absolutely necessary for K_V [5], [9]. A detailed discussion of the aforementioned parameters can be found in [5]. Table 1 summarizes basic parameters and the Euler number density χ_V of both microstructures.

Table I. Basic parameters (V_V, S_V, M_V and K_V) and Euler number density (χ_V) for oxides and pores in Ag/SnO_2 contact materials.

	Initial state		Eroded	
	oxides	Pores	oxides	Pores
V_V, %	20.4	0.33	21.5	3.88
S_V, μm^{-1}	1.12	0.066	1.25	0.289
M_V, μm^{-2}	2.08	0.497	1.40	0.839
K_V, μm^{-3}	0.44	4.876	-3.48	2.912
χ_V, μm^{-3}	0.03	0.388	-0.28	0.232

In the electro eroded state, volume fraction of the SnO_2 particles is approx. 1 % higher than in the initial state. Reconstruction has been limited to an area with oxide agglomerates around the pores and thus volume fraction of pores is ten times higher. This is the first indication for the local inhomogenity and the degradation caused by the electro erosion process. Specific surface area is higher in case of the eroded sample. Because of the similar volume fractions, this means, that particles have more complex shape rather than spherical one. The integrals of mean curvature and total curvature, which describe the dispersion of the microstructure, are higher for the initial state. These values are connected to the number of particles in case of a homogeneous microstructure. For disperse microstructures, e.g. Ag/SnO$_2$ in the initial state, particle density (N_V) and Euler number density (χ_V) are equal. So particle density can be determined from the basic parameters of the microstructure. In case of complex, interconnected microstructures, e.g. the switched sample, particle number doesn't provide significant results. In a complete network, particle number is only 1. In these cases, Euler number density, which considers additionally the connectivity of particles has more importance [5].

The 3D quantitative image analysis software MAVI allows to determine statistical values of particle based parameters like size and shape. The agglomerates of oxides can be analyzed not only as a whole part, but also as separated objects. For that, agglomerates have to be separated using the Euklidian distance transformation (EDT) and the watershed transformation. Particles in the initial state are smaller (d = 1.97 ± 0.03 µm) and more spherical (f = 0.59 ± 0.17) than the oxides in the eroded sample (d = 2.68 ± 0.05 µm, f = 0.47 ± 0.25). f is the ratio between maximum and minimum diameter, so an ideal sphere has a value of f=1. Figure 3 shows the distribution of particle number and volume fraction for different size classes.

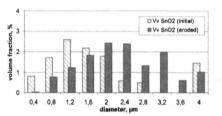 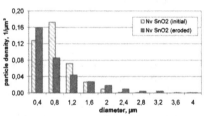

Figure 3. Particle density and volume fraction for different size ranges. Size distribution is shifted towards bigger particles in case of the eroded sample.

Not only size and shape, but also connectivity and distance between the particles influence considerably the properties of the contact material.

Distance distribution of oxides

The oxides in the initial state are mainly separated from each other (dispersed). Oxides in the eroded state are generally concentrated around pores. For quantitative understanding of the distance distribution of the oxides, morphological operations or the EDT may be used. The morphological operation "dilation" adds a certain number of voxels to the particles, so the particles continuously grow in all spatial directions. After every step of dilation, particle number and increase of the particle volume is recorded (figure 4). After a certain number of steps, only one particle, which fills the whole volume, is present [9].

With the EDT, the distance of each voxel of the matrix to the surface of the oxides is calculated and related to a certain grayscale. Figure 4 shows the distribution of oxide free areas as a function of the distance between the oxides. The curves allow us to describe the evident differences quantitatively. A maximum at small values indicates the clustering. The slow increase of the volume fraction indicates bigger oxide free parts in the eroded sample. For the initial state, a local maximum at 250 to 500 nm can be seen, so dispersion in the matrix is more homogeneous. Distance between the oxides in the initial state is smaller than in the eroded state.

The EDT method is suited especially for complex microstructures with particles of different shape, size and orientation. When the connectivity is considered, EDT method provides comprehensive and statistically secured quantification for homogeneous and inhomogeneous microstructures.

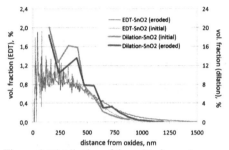

Figure 4. Distribution of the oxide free volume as a function of distance to the surface of the oxides, determined by EDT and dilation. Most of oxides in the eroded sample are located in vicinity of other oxides (< 100 nm).

3D simulation of effective properties

The effective electrical conductivity has been acquired using the simulation software GeoDict on real tomographic data. High resolution FIB-nanotomography provides a very large number of voxels (6.5×10^7) and thus simulation times are very long. It has been shown, that coarsening of the tomographic data up to 120 nm voxel size has only a small effect on the simulation results but leads to considerably shorter simulation times.

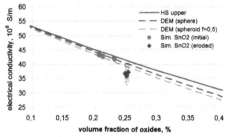

Figure 5. Comparison of simulated effective properties and analytical models (Hashin Shtrikman bounds HS and differential effective medium approach DEM). Simulated conductivity corresponds to the analytical models for the initial state but not for the eroded state.

Concluding, the optimization of the volume for the simulation can continue as long as the voxel size does not reach the critical microstructure parameters (here particle size d $\approx$ 2 μm and particle distance approx. 0.5 – 1 μm) or the basic field features (V_V, S_V, M_V, K_V) change.

Simulation results for the initial state agree with the analytical models (Hashin Shtrikman bounds (HS) and differential effective medium approach (DEM) [10]) (figure 5). Electrical conductivity in the eroded sample is lower due to the high connectivity of the oxide agglomerates.

CONCLUSIONS

Two different microstructures of Ag/SnO$_2$ contact materials have been analyzed using FIB-nanotomography and quantitatively characterized with 3D image analysis. Euklidian distance transformation (EDT) was applied to describe the distributions of oxides in silver-matrix. This way it was possible to quantify the agglomeration of the oxide particles and microstructure degradation in the eroded sample.

It was shown that the simulation process can been optimized by coarsening of the tomographic data. Voxel size can be increased until the critical size of the microstructure components is reached or the basic parameters of the microstructure change.

Comparison of simulation results on the real tomographic data and the analytical models for prediction of effective properties shows, that connectivity, which can only be determined from 3D data, plays an important role for the materials properties.

Thus, the combination of FIB-nanotomography with modern 3D image analysis and simulation methods offers new possibilities in target characterization of contact materials.

ACKNOWLEDGMENTS

This work was supported by the BMBF-Project No. 03X3500.

REFERENCES

[1] Jeanvoine, N., et al. *Pract. Metallography.* 2006, Vol. 43, pp. 107-119.
[2] Soldera, F., et al. *Microscopy and Microanalysis.* 2007, Vol. 13, 3, pp. 422-423.
[3] Lasagni, F., et al. *Adv. Eng. Mat.* 2008, Vol. 10, pp. 62-66.
[4] Konrad, J., Zaefferer, S. and Raabe, D. *Acta Mater.* 2006, Vol. 54, 5, pp. 1369-1380.
[5] Ohser, J. and Mücklich, F. *Statistical Analysis of Microstructures in Materials Science.* s.l. : John Willey & Sons, 2000.
[6] Ohser, J., Nagel, W. and Schladitz, K. *Image Anal. Stereol.* 2003, Vol. 22, pp. 11-19.
[7] Velichko, A., et al. *Acta Mater.* 2008, Vol. 56, pp. 1981-1990.
[8] Wiegmann, A. and Zemitis, A. EJ-HEAT,Bericht des Fraunhofer ITWM, Nr. 94, 2006.
[9] Velichko, A. and Mücklich, F. *Pract. Metallography.* 2008, Vol. 45, pp. 423-439.
[10] Weber, L, Dorn, J. and Mortensen, A. *Acta Mater.* 2003, Vol. 51, pp. 3199-3211.

Mater. Res. Soc. Symp. Proc. Vol. 1184 © 2009 Materials Research Society   1184-HH03-07

Study of Electron Beam Irradiation Induced Defectivity in Mono and Bi Layer Graphene and the Influence on Raman Band Position and Line-Width

G. Rao[1], S. Mctaggart[1], J. L. Lee[1], and R. E. Geer[1]
[1]College of Nanoscale Science and Engineering, University at Albany, SUNY, Albany NY 12222, USA

ABSTRACT

Nanoscale metrology of graphene-based devices is a substantial challenge. The investigation of defects and stacking order is essential for graphene-based device development. Raman spectroscopy is a useful approach in this regard. The defect-induced Raman D band yields substantial insights regarding defect density and, consequently, can serve as in important tool to quantify impact of defects on eventual graphene-based device performance. Toward this end an investigation of electron beam-induced defects in bi-layer and mono layer graphene samples has been undertaken via the examination of the Raman D, and G bands. The evolution of the aforementioned Raman spectra as a function of electron beam dose was characterized via Raman spectroscopy and compared with spectra from the same samples prior to irradiation. Defect generation in the graphene as a function of electron beam dose was characterized via the change in the intensity ratios of the Raman D and G bands (I_D/I_G) and the broadening of the G band line width. Continued irradiation at very high flux and very low accelerating voltages have also revealed charge accumulation evident from the narrowing of G band line-widths.

INTRODUCTION

Graphene, a two-dimensional hexagonal, sp^2 coordinated array of carbon atoms provides new possibilities for post-CMOS electronics[1,2]. Graphene exhibits remarkable electronic properties including ballistic carrier transport with room temperature carrier mobilities as high as 20 000 cm^2/Vs[1,3]. The linear electronic dispersion of graphene is postulated to imbue charge carrier behavior akin to Dirac-Fermions[4]. However, the presence of edge defects and disorder in the material has been shown to degrade these properties. It has been reported that edge defects and disorder reduces the mobility to 3000-5000 cm^2/Vs[5].

Since defects and lattice distortion in graphene can be introduced via energetic charged particles it is essential to characterize the effects of characterization or device fabrication tools employing such (e.g. SEM, TEM , FIB EB Litho, etc…). Defect formation has been observed earlier in carbon nanotubes with electron irradiation in a SEM. Very low accelerating voltages (~0.5, 1kV) have proved to induce more extensive damage[6]. The induced defectivity increased electron scattering in the CNT material thereby reducing overall electron mobility[7]. Consequently, the work presented here has focused on the effects of very low energy electron irradiation on mono (1L) and bi (2L) layer graphene samples. It is observed that electron irradiation results in increased defect formation in both 1L and 2L graphene per the variation of the Raman D/G intensity ratio. Considering the energy spectrum of the radiation, the increased defect signature is attributed to carbon-carbon bond-length variations (disclination-type defects) as opposed to beam-induced vacancies or interstitials[8]. Increased irradiation at very low accelerating voltages resulted in modification of the Raman G band consistent with electrostatic doping of the graphene induced by charge accumulation at the sample surface. The unique signatures of Raman scattering associated with such lattice modifications underscores its utility in studying and characterizing defects in graphene and other carbon-based materials of interest for post-CMOS nanoelectronics[9].

SAMPLE PREPARATION AND EXPERIMENTAL SET-UP

Samples were prepared by mechanical exfoliation of highly oriented pyrolitic graphite grade ZYA on 300nm SiO_2/Si substrates[1,10]. The 1L and 2L graphene exfoliates were identified initially by optical microscopy. The monolayer or registered bilayer character of a given graphene exfoliate was confirmed by analyzing the Raman 2D band. Monolayer exfoliates possessed a single symmetric 2D band shifted to lower wave number compared to the graphite 2D Raman band. Registered bilayer graphene exfoliates exhibited the characteristic four fold splitting of the 2D band as previously reported[11]. A Renishaw R100 Raman spectrometer with a 1800 lines/mm grating was used with a 514.5 nm laser excitation wavelength. In order to avoid sample damage the incident laser power at the sample was limited to 200µW with a spot size diameter of 0.8µm. All measurements were taken with a 50X objection lens (N.A of 0.75). All Raman measurements were taken immediately before and after e-beam irradiation.

EXPERIMENTAL DESIGN

All e-beam irradiation of graphene samples were performed at 0.5 keV accelerating voltage using a LEO 1550 SEM . The irradiation dose was determined for a given exposure time using a Faraday cup. The measured e-beam current at 0.5 keV accelerating voltage was 15.6 pA. The calculated doses for continuous irradiation and sequential irradiation are listed in Table I.

Table I. Flux variation for continuous irradiation

Time (minutes) Continuous Irradiation	Dose (cm^{-2})	Time (minutes) Sequential Irradiation	Dose (cm^{-2})
1	13.68 x 10^{15}	1	13.68 x 10^{15}
3	41.04 x 10^{15}	4	54.72 x 10^{15}
5	68.4 x 10^{15}	9	123.12 x 10^{15}
7	95.76 x 10^{15}	16	218.88 x 10^{15}

Samples that underwent 'continuous irradiation' remained in the SEM sample chamber under vacuum for the entirety of the e-beam exposure (i.e. a given dose). Samples that underwent 'sequential irradiation' were removed from the SEM vacuum chamber after each exposure cycle listed in Table I. Raman characterization was carried out after each exposure. This approach was taken to better simulate the effect of air (oxygen, humidity) exposure as would be encountered in conventional multi-step processing.

152

Continuous radiation doses ranged from 13.7×10^{15} cm^{-2} to 95.8×10^{15} cm^{-2}. Sequential radiation doses ranged from 13.7×10^{15} cm^{-2} to 218.9×10^{15} cm^{-2}.

RESULTS AND DISCUSSIONS

The Raman spectra from a 1L graphene exfoliate (pre- and post-irradiation) is plotted in Fig. 1 in the vicinity of the D and G bands. In an ideal 1L graphene lattice (free from disorder) D-band Raman scattering is not expected as it results from disorder-induced scattering near the K-point in the Brillouin zone via the mechanism of resonance Raman scattering[9]. This is the case for the pre-irradition graphene data for which D band scattering is essentially absent. However, a substantial D band Raman peak appears on exposure to an electron beam (Fig. 1). The I_D/I_G ratios for 1L and 2L graphene exfoliates are plotted in Fig. 2a (continuous irradiation) and in Fig. 2b (sequential irradiation).

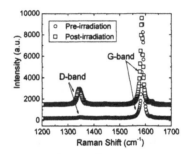

Figure 1 Raman spectra from a 1L graphene exfoliate (pre- and post-irradiation). Appearance of D-band is indicative of disorder/defect generation.

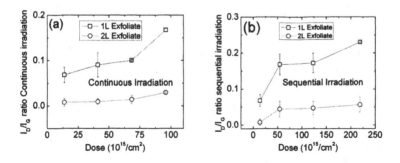

Figure 2 (a) I_D/I_G ratio for continuous irradiation. (b) I_D/I_G ratio for sequential irradiation.

For the continuous irradiation experiments four different 1L and 2L flakes were irradiated independently at the doses listed in Table I. The I_D/I_G ratios (Fig. 2a) increase with increasing electron irradiation dose. Note the I_D/I_G ratios for 1L exfoliates are higher than the 2L samples as expected. The total areal atomic density for a 2L exfoliate is twice that for a 1L sample and reduces the net flux per atom in the bi-layer lattice. The I_D/I_G ratios for the corresponding sequentially-irradiation graphene samples likewise exhibit an increase with dose, although this increase is substantially larger for a given dose following ambient exposure. This is attributed to ambient exposure (e.g. oxidation) effects. The I_D/I_G ratios of 1L exfoliates for sequential radiation are also larger compared to those of 2L exfoliates for the same reasons as noted above.

Electron irradiation also results in substantial G band broadening in both 1L and 2L graphene exfoliates. This is illustrated in Fig.3 for continuously irradiated samples.

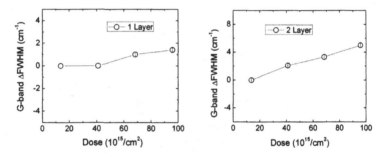

Figure 3. Broadening of G band FWHM on continuous irradiation in 1L and 2L graphene. The increase in line width with dose is consistent with increased disorder and reduced phonon lifetimes. In both 3(a) and (b)The error bars are less than the symbol size

The broadening of the G bands with e-beam irradiation implies increased electron-phonon coupling (EPC) and reduced phonon lifetimes[7]. This is clearly due to the increased graphene lattice defect density indicated by the non-zero I_D/I_G ratios (Fig. 1) which results in increased electron-phonon scattering[9,12].

Although the Raman spectra data in Figs. 1 and 2 clearly indicate increased disorder or defectivity, the Raman bands, themselves, are not defect-specific. However, the probability of irradiation-induced vacancy or interstitial formation in the 1L and 2L graphene exfoliates is very low taking into account the accelerating potentials used. The electron beam energy required for ejection of an atom from the lattice is 86keV[6,13]. The electron beam accelerating voltage for irradiation utilized for this work was 0.5kV. Consequently, the irradiation-induced disorder is predominantly due to bond length variations occurring in the graphene lattice. Such bond length variations would ostensibly be reflected in the formation of Shuffle or Stone Wales defects[14].

In contrast to the data presented in Fig 3, 1L and 2L graphene exfoliates subjected to sequential irradiation exhibited a reduction in the FWHM observed as shown in Fig 4. Specifically, the 1L and 2L exfoliates exhibit reductions in the G band width of roughly 1 cm^{-1} and 5 cm^{-1}, respectively. Reductions in G band line widths have been reported to result from doping or charge accumulation[15]. From our experiments it is clear that the G band line width reduction is mediated or enhanced by ambient exposure. This, in turn, implies that the doping or charge accumulation is tied to local oxidation (or contamination) promoted through irradiation-induced bond length disorder.

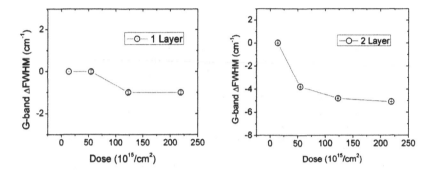

Figure 4 Reduction in FWHM of G peak on sequential irradiation in (a) 1L and (b) 2L graphene exfoliates. In both cases 4(a) and (b) error bars are less than the symbol size.

It is interesting to note that the reduction in the G band line width observed for sequential irradiation is substantially larger for the 2L graphene exfoliate compared to the 1L exfoliate. This is somewhat counter intuitive based on the I_D/I_G ratio data in Fig. 2b which may be taken to indicate relatively higher defectivity in the monolayer exfoliates. However, it is consistent with a scenario whereby local ambient oxidation (upon sample exposure to air) is sensitive to local changes in carbon-bond hybridization from sp^2 to sp^3 which may be substantially more prevalent in bilayer graphene as opposed to monolayer graphene.

CONCLUSIONS

It has been shown that very low energy electron beams are capable of inducing substantial disorder in graphene as confirmed by Raman spectroscopy. Defect generation in the graphene as a function of electron beam dose was characterized via the change in the intensity ratios of the Raman D and G bands (I_D/I_G) and the broadening of the G band line width. Moreover, the combination of repeated electron-irradiation with ambient exposure has been shown to substantially alter the quantitative and qualitative nature of the induced defectivity. Since characterization, metrology, and process tools conventionally utilized for nanoelectronic device fabrication employ focused electron or ion beams it is essential to carefully characterize resultant impacts on monolayer and bilayer graphene if these materials are to be effectively used for post-CMOS nanoelectronics.

REFERENCES

[1]   K. S. Novoselov, A. K. Geim, S. V. Morozov, D. Jiang, Y. Zhang, S. V. Dubonos, I. V. Grigorieva, and A. A. Firsov, Science **306,** 666-669 (2004).

[2]   A. K. Geim and K. S. Novoselov, Nature Materials **6,** 183-191 (2007).

[3]   J. C. Charlier, P. C. Eklund, J. Zhu, and A. C. Ferrari, Carbon Nanotubes **111,** 673-709 (2008).

[4]   Y. B. Zhang, Y. W. Tan, H. L. Stormer, and P. Kim, Nature **438,** 201-204 (2005).

[5]   L. Tapasztó, G. Dobrik, P. Nemes-Incze, G. Vertesy, P. Lambin, and L. P. Biro, Physical Review B **78,** 233407 (2008).

[6]   K. K. S.Suzuki , Y. Homma, and S.-y. Fukuba, Japanese Journal of Applied Physics **43,** L1118-L1120.

[7]   M. Lazzeri, S. Piscanec, F. Mauri, A. C. Ferrari, and J. Robertson, Physical Review B (Condensed Matter and Materials Physics) **73,** 155426-6 (2006).

[8]   D. Teweldebrhan and A. A. Balandin, Applied Physics Letters **94,** 013101 (2009).

[9]   M. A. Pimenta, G. Dresselhaus, M. S. Dresselhaus, L. G. Cancado, A. Jorio, and R. Saito, Physical Chemistry Chemical Physics **9,** 1276-1291 (2007).

[10]   K. S. Novoselov, D. Jiang, F. Schedin, T. J. Booth, V. V. Khotkevich, S. V. Morozov, and A. K. Geim, Proceedings of the National Academy of Sciences of the United States of America **102,** 10451-10453 (2005).

[11]   A. C. Ferrari, J. C. Meyer, V. Scardaci, C. Casiraghi, M. Lazzeri, F. Mauri, S. Piscanec, D. Jiang, K. S. Novoselov, S. Roth, and A. K. Geim, Phys Rev Lett **97,** 187401 (2006).

[12]   A. C. Ferrari, Solid State Communications **143,** 47-57 (2007).

[13]   C. O. Girit, J. C. Meyer, R. Erni, M. D. Rossell, C. Kisielowski, L. Yang, C.-H. Park, M. F. Crommie, M. L. Cohen, S. G. Louie, and A. Zettl, Science **323,** 1705-1708 (2009).

[14]   A.Carpio, Physical Review B **78** (2008).

[15]   C. Casiraghi, S. Pisana, K. S. Novoselov, A. K. Geim, and A. C. Ferrari, Applied Physics Letters **91** (2007).

Mater. Res. Soc. Symp. Proc. Vol. 1184 © 2009 Materials Research Society   1184-HH04-08

Structural Characterization of GeSn Alloy Nanocrystals Embedded in SiO$_2$

Swanee J. Shin,[1,2] Julian Guzman,[1,2] Chun-Wei Yuan,[1,2] Christopher Y. Liao,[1,2] Peter R. Stone,[1,2] Oscar D. Dubon,[1,2] Andrew M. Minor,[1,2] Masashi Watanabe,[2] Joel W. Ager III,[2] Daryl C. Chrzan,[1,2] and Eugene E. Haller[1,2]

[1]Department of Materials Science and Engineering, University of California, Berkeley, CA 94720, USA
[2]Materials Sciences Division, Lawrence Berkeley National Laboratory, Berkeley, CA 94720, USA

ABSTRACT

GeSn alloy nanocrystals were formed by implantation of Ge and Sn ions into an amorphous SiO$_2$ matrix and subsequent thermal annealing. High resolution transmission electron microscopy (HRTEM) and scanning transmission electron microscopy (STEM) with a high angle annular dark field (HAADF) detector were used to show that phase-segregated crystalline bi-lobe nanocrystals were formed. Rapid melting and solidification using a single excimer laser pulse transformed the bi-lobe structure into a homogeneously mixed amorphous structure. Raman spectroscopy was used to monitor the crystalline nature and approximate grain size of the Ge portion of the nanocrystals after each heat treatment, and the Raman spectra were compared with the TEM images.

INTRODUCTION

Nanocrystals have attracted considerable attention due to their unique size dependent properties and wide range of potential applications in integrated electronic devices, optoelectronic devices, and energy generation systems. The ion beam synthesis (IBS) method has been used to synthesize nanocrystals for many years.[1] Recently, our work has focused on understanding the role of the matrix/nanocrystal interface in determining the properties of nanocrystals made by IBS[2]; for example, we found that the melting point of Ge nanocrystals embedded in silica can be elevated by more than 200°C above the bulk value.[3] Two or more elements can be used to make alloy or compound nanocrystals.[1,4] In some cases, there are multiple phases and geometries possible, which are determined by the significant surface and interface effects at this nanoscale.[5,6] In particular, Ge$_{1-x}$Sn$_x$ alloys have been intensively studied as a promising material for light emitting devices due to the reported indirect to direct bandgap transition at x≈0.1[7-9], and at the nanoscale, GeSn quantum dots showed the size dependent quantum confinement effect.[10,11] Several nonequilibrium growth methods have been attempted for films[8,9] and quantum dots[10,11], most of which used epitaxial growth on the underlying substrate. In this study, we used the IBS and laser processing method to synthesize GeSn nanocrystals embedded in an amorphous matrix, and several techniques were used to characterize fully the structure of the nanocrystals, and complementary information each method provided will be highlighted.

EXPERIMENT

Isotopically pure ^{74}Ge and ^{120}Sn were implanted at room temperature into 500 nm thick amorphous SiO_2 layers grown on Si(100) substrate by wet oxidation. The energy and dose of ^{74}Ge and ^{120}Sn were selected to be 150 keV at $4\times10^{16}cm^{-2}$ and 120 keV at $1\times10^{16}cm^{-2}$, respectively, such that the peak positions of implantation profiles could approximately match each other. Subsequent annealing was performed in a sealed ampoule at 900°C for 1 hr under Ar atmosphere, and the samples were quenched under cold running water.

For the pulsed laser melting (PLM) process, an annealed sample was irradiated with a single 248 nm KrF excimer laser pulse. The pulse had 30 ns duration time with 0.3 J/cm^2 energy fluence. The laser fluence was empirically chosen such that the maximum temperature exceeds the melting point of the nanocrystals and the heat dissipation from the given sample geometry was assumed to be fast enough to quench the nanocrystals in their supercooled liquid state. To detect any change from the original elemental distribution, a Rutherford backscattering spectrometry (RBS) measurement was performed between each process.

To investigate the size distribution and morphology of the nanocrystals, transmission electron microscopy (TEM) was used. To distinguish between Ge and Sn, Z-contrast imaging using a scanning transmission electron microscope (STEM) with a high angle annular dark field (HAADF) detector was performed. Also, HRTEM was used on single nanocrystals to confirm the crystallinity of both the Ge and the Sn phase and to investigate any potential defects at the interfaces. Raman spectroscopy with 488nm Ar$^+$ laser line was used to monitor the crystallinity of Ge phase and the size dependent phonon confinement.

RESULTS AND DISCUSSION

Microscopic local probe (HAADF-STEM)

Figure 1(a) shows a HAADF-STEM image of the nanocrystals after thermal annealing. They are clearly phase-separated "bi-lobe" nanocrystals, and the diameter of individual nanocrystal is approximately 10~20 nm. Note that there exists a depth dependent atomic concentration distribution from the surface as shown in Figure 1(b). The size and composition of each nanocrystal directly reflect this profile, and the largest nanocrystals are formed at ~80nm depth from the surface as shown in Figure 1(a). Since HAADF-STEM image contrast is sensitive to the atomic number Z, independent of the crystallographic orientation, and Ge/Sn system is hardly miscible, we can readily know that bright regions correspond to the heavier atom (Sn) rich phase while the dark regions are the lighter atom (Ge) rich phase. The nanocrystals are randomly oriented because they are embedded in an amorphous matrix, making it difficult to quantify the volume fraction of each phase. However, the Ge to Sn concentration ratio in Figure 1(b) near the peak depth (~80 nm) and the average Ge-rich to Sn-rich volume extracted from many HAADF-STEM images are in reasonable agreement. (~2:1)

Figure 1(c) is an image taken after the PLM process of the bi-lobe nanocrystals. The Z-dependent contrast across individual nanocrystal has disappeared while the size distribution remained unchanged. To ensure Ge and Sn atoms are retained after PLM and to monitor any change in distribution profile, an RBS measurement was performed. Figure 1(b) indicates that the total atomic concentration and distribution have little difference before and after PLM, which means both elements are contained in the nanocrystals, and they are a homogeneously mixed alloy. According to the bulk Ge-Sn phase diagram, Ge and Sn are hardly miscible through the entire composition range, and it can be inferred from Fig 1(a) that given enough time (i.e. one hour at 900°C), GeSn forms bi-lobe nanocrystals as the equilibrium configuration. Therefore, it can be concluded that homogeneously mixed nanocrystals are in a kinetically limited metastable state due to the much faster cooling rate of the PLM process compared to furnace annealing.

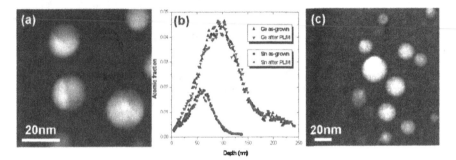

Figure 1. (a) HAADF-STEM image of GeSn nanocrystals after annealing at 900°C for 1hr. Phase separated bi-lobe nanocrystals are formed. (b) RBS spectra of implanted Ge and Sn atoms before and after PLM process. (c) HAADF-STEM image after PLM of (a). Homogeneously mixed alloy nanocrystals are formed.

Crystallinity measurements (Raman spectroscopy + HRTEM)

Raman spectra from samples before and after PLM are shown in Figure 2(a). Bi-lobe nanocrystals show a strong crystalline Ge peak at ~300 cm^{-1}, and the peak is asymmetrically broadened to the lower energy side due to the phonon confinement effect. The shape and width of the spectrum is identical to that of pure Ge nanocrystals formed with the same implantation dose and energy, thus the size distribution of Ge phase out of GeSn bi-lobe nanocrystals are thought to be the same as pure Ge nanocrystals.[2] In contrast, after the PLM process, the crystalline Ge peak is significantly suppressed. Instead, the spectrum shows a broad peak at ~275 cm^{-1}. This feature has been observed both from sputtered Ge film and free-standing Ge nanocrystals exposed in air, and it is attributed to disordered Ge-Ge bonds from the amorphous phase.[12] This fact further supports our contention that the GeSn nanocrystals after PLM are in an amorphous state.

To further investigate the crystallinity, HRTEM images were taken from bi-lobe nanocrystals. Figures 2(b) and (c) show images from a single nanocrystal taken at slightly different tilt angles to find a zone axis perpendicular to a lattice plane of each phase. From the images, each phase is found to be single crystalline, and no defects could be found. Measured interplanar distances of Figure 2(b) (top right phase) and (c) (bottom left phase) are 3.24 Å and 2.00 Å, which correspond to Ge (111) and β-Sn (211) planes, respectively. This measurement confirms that each phase consists of nearly pure Ge or Sn. It appears that the two phases are oriented to cause minimum lattice mismatch at the interface, and thus form defect free single crystalline phases on both sides.

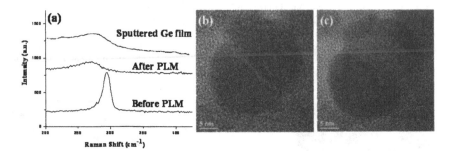

Figure 2. (a) Raman spectra from GeSn nanocrystals before PLM (bi-lobe), after PLM (homogeneous), and sputtered Ge film (amorphous). (b), (c) HRTEM images of a bi-lobe nanocrystal. Lattice planes are visible from (b) top-right phase and (c) bottom-left phase.

CONCLUSIONS

GeSn alloy nanocrystals were formed in an amorphous SiO_2 matrix by the IBS method and PLM process, and several complementary techniques were applied to fully characterize them. HAADF-STEM images showed they are bi-lobe and mixed alloy structure before and after PLM, respectively. Raman spectroscopy and HRTEM measurements were performed to investigate crystallinity, and revealed that bi-lobe nanocrystals consist of phase separated defect free single crystalline Ge and Sn, while mixed alloy is in an amorphous state.

ACKNOWLEDGMENTS

This work was supported in part by the Director, Office of Science, Office of Basic Energy Sciences, Division of Materials Sciences and Engineering, of the U.S. Department of Energy under Contract No. DE-AC02-05CH11231 and in part by U.S. NSF Grant Nos. DMR-0405472. Portions of this research were conducted at the National Center for Electron Microscopy, Lawrence Berkeley National Laboratory, which is supported by the U.S. Department of Energy under Contract No. DE-AC02-05CH11231.

REFERENCES

1. A. Meldrum, R. F. Haglund, L. A. Boatner, and C. W. White, *Adv. Mater.* **13**, 1431 (2001).
2. I. D. Sharp, Q. Xu, C. Y. Liao, D. O. Yi, J. W. Beeman, Z. Liliental-Weber, K. M. Yu, D. N. Zakharov, J. W. Ager, III, D. C. Chrzan, and E. E. Haller, *J. Appl. Phys.* **97**, 124316 (2005).
3. Q. Xu, I. D. Sharp, C. W. Yuan, D. O. Yi, C. Y. Liao, A. M. Glaeser, A. M. Minor, J. W. Beeman, M. C. Ridgway, P. Kluth, J. W. A. III, D. C. Chrzan, and E. E. Haller, *Phys. Rev. Lett.* **97**, 155701 (2006).
4. P. Kluth, B. Hoy, B. Johannessen, S. G. Dunn, G. J. Foran, and M. C. Ridgway, *Appl. Phys. Lett.* **89**, 153118 (2006).

5. M. Strobel, K.-H. Heinig, and W. Moller, *Nucl. Instrum. Methods Phys. Res., Sect. B* **148**, 104 (1999).
6. C. W. Yuan, S. J. Shin, C. Y. Liao, J. Guzman, P. R. Stone, M. Watanabe, J. W. Ager, III, E. E. Haller, and D. C. Chrzan, *Appl. Phys. Lett.* **93**, 193114 (2008).
7. D. W. Jenkins and J. D. Dow, *Phys. Rev. B* **36**, 7994 (1987).
8. G. He and H. A. Atwater, *Phys. Rev. Lett.* **79**, 1937 (1997).
9. V. R. D'Costa, C. S. Cook, A. G. Birdwell, C. L. Littler, M. Canonico, S. Zollner, J. Kouvetakis, and J. Menendez, *Phys. Rev. B* **73**, 125207 (2006).
10. Y. Nakamura, A. Masada, and M. Ichikawa, *Appl. Phys. Lett.* **91**, 013109 (2007).
11. N. Naruse, Y. Mera, Y. Nakamura, M. Ichikawa, and K. Maeda, *Appl. Phys. Lett.* **94**, 093104 (2009).
12. A. J. Williamson, C. Bostedt, T. vanBuuren, T. M. Willey, L. J. Terminello, G. Galli, and L. Pizzagalli, *Nano Lett.* **4**, 1041 (2004).

Mater. Res. Soc. Symp. Proc. Vol. 1184 © 2009 Materials Research Society   1184-HH02-02

Structure of Cleaved (001) USb₂ Single Crystal

Shao-Ping Chen[1], Marilyn Hawley[1], Phil B. Van Stockum[2], Hari C. Manoharan[2] and Eric D. Bauer[1]

[1]Los Alamos National Laboratory, Los Alamos, New Mexico; [2]Department of Physics and Stanford Institute for Materials and Energy Sciences, Stanford University, Stanford, California

ABSTRACT

We have achieved what we believe to be the first atomic resolution STM images for a uranium compound taken at room temperature. The a, b, and c lattice parameters in the images confirm that the USb2 crystals cleave on the (001) basal plane as expected. The a and b dimensions were equal, with the atoms arranged in a cubic pattern. Our calculations indicate a symmetric cut between Sb planes to be the most favorable cleavage plane and U atoms to be responsible for most of the DOS measured by STM. Some strange features observed in the STM will be discussed in conjunction with ab initio calculations.

INTRODUCTION

The purpose of this work is to demonstrate the power of scanning tunneling microscopy (STM) techniques combined with a theoretical underpinning to determine the surface atomic structure and properties of actinide materials, such as the quasi 2-dimensional uranium dipnictide USb$_2$ single crystal, thereby contributing to the understanding of their surface structural and electronic properties. The members of this interesting UX$_2$ (X=P, As, Sb, Bi) series of compounds display dual localized and itinerant 5f electron behavior within the same compound due to the hybridization of the 5f orbitals with the conduction band [1]. With the exception of UO$_2$, which has to be studied at elevated temperature to generate enough carriers for STM imaging [2,3,4], STM techniques have not been applied successfully to the characterization of the surface atomic structure of any other single crystal actinide compound, to the best of our knowledge. However, STM has been used to a limited extent for the study of some cerium compounds [5]. STM probes electronic properties at the atomic level and can directly provide information about the local density of filled and empty states (LDOS) states simultaneously. A STM topograph provides the local atomic arrangement and spacing of the atoms on the surface, local defect structures (e.g. steps, vacancies, and kink sites), and the presence of contaminants, all of which are averaged over when probed in photoemission studies.

The quasi two-dimensional USb$_2$ has a layered tetragonal structure that is easily cleaved and has been extensively studied by a number of different techniques, such as resistivity [6], Hall effect measurements [7], photoemission [8] and angle-resolved photoemission spectroscopy [9,10], de Haas-van Alphen [11-13], neutron diffraction [14], nuclear magnetic resonance [15], and U^{238} Mossbauer spectroscopy [16] techniques. Here, we provide local information about the surfaces of this interesting compound, which we find to contain a high density of defects.

EXPERIMENT

Single crystals of USb$_2$ were grown in Sb flux growth [17]. The crystals were prepared for STM imaging using the following procedure: the crystals were cleaved between a-b planes to obtain thin flat crystals, which were then attached to STM platens using conductive silver epoxy; a small piece of sapphire substrate was epoxied to the top surface of each crystal using Torrseal™; and a short titanium post was attached to the sapphire substrate to facilitate in situ UHV cleaving. The freshly cleaved surfaces were characterized using an Omicron UHV-STM operating under a pressure of about 5×10^{-10} Torr. Imaging was carried out using cut Pt-Ir tips under typical tunneling conditions of ±1V sample bias voltage and 125 pA tunneling current. Both filled and empty state images were taken on one of the samples to help understand the origin of the observed surface features. A freshly prepared pristine Si(001)-2x1 surface was used to calibrate the piezoelectric scanner to correct the observed atomic spacing for the USb$_2$ crystals. All measurements were conducted at room temperature.

USb$_2$ has a PbFCl or anti-Cu$_2$Sb (P4/nmm) structure with a = b = 0.4270 nm, c = 0.8748 nm [10] and c/a = 2.049. In this tetragonal layered structure, it is expected to cleave between a-b planes, which have a square arrangement of atoms in the plane perpendicular to the long axis along [001]. Figure 1 shows STM image collected from one of the samples.

Fig. 1

The STM images reveal that the atoms on the cleaved surface are arranged in a square pattern, as expected for the sample cleaving between a-b basal planes. The measured distance between rows of atoms in the plane, which has been corrected using measurements from a Si(100)-2x1 surface, is 0.441±0.006 nm. This value is slightly larger than the published bulk value mentioned above of 0.4270 nm [10].

Although the square atomic arrangement is clearly visible, the surface contains a significant number of what appear to be missing atoms, primarily in rows corresponding to the <100> and <010> crystal directions with equal probability. The most common feature is a single atom vacancy, followed by two to three adjacent vacancies. In order to distinguish vacancies from the presence of another type of atom (e.g. contaminant or oxygen) whose DOS differs significantly from that of the majority of atoms on the surface (in this case either U or Sb), filled and empty state images were taken simultaneously. The filled and empty state images are nearly indistinguishable, suggesting that the darker features are most likely vacancies rather than differences in the DOS between atomic species. Even though the individual atoms are not as clearly visible as in figure 1, the rows of atoms and "vacancies" are aligned at approximately 90° angles, and the atomic spacing is the same as in the first sample. Unlike the crystal shown in figure 1, this area of the second sample had a nearly equal number of atoms and "vacancies" The topographic height difference between the surface atoms and the bottom of the dark features in both sets of images is 0.080±0.014 nm, which is discussed below.

CALCULATIONS

We have used ab initio density functional theory (DFT) [18] to study the surface energies of various terminations of the (001) surface of USb_2. The projector augmented wave method combined with the generalized gradient approximation [19] was used to describe the U and Sb. The k-space sampling was done through the Monkhorst-Pack scheme. Valence electrons in both U and Sb atoms are treated as itinerant. The convergence of the calculations was checked to ensure an energy convergence of better than 1 meV/atom and forces less than 0.5×10^{-3} eV/ Å .

USb_2 crystallizes in the PbFCl (P4/nmm) structure with U in the (0.25, 0.25, u) and Sb in the (0.75, 0.75, v) position, where $u = 0.280$ and $v = 0.365$ [14]. The calculated value for a is 0.4274 nm while holding the value of c/a, u, and v fixed at the experimental values. If we allow u and v to change while fixing the c/a ratio, the relaxed value for u is 0.279, which is almost identical to the experimental value of 0.280. The relaxed v value is 0.359, which is again very close to the experimental value of 0.365. Overall, the agreement with experiment is very good. The anti-ferromagnetic calculations of the bulk do confirm the magnetic moment arrangement that has been observed in other experiments [15]. There are no significant differences in the energy and structural properties between the results of anti-ferromagnetic and nonmagnetic calculations. There are many possible configurations for the magnetic moments in the actinides [20]. Here we only consider nonmagnetic, ferromagnetic and antiferro magnetic solutions. We also performed relativistic calculations with spin-orbit interactions explicitly included and found that the cleavage energies and structural properties are only slightly modified and the results for the fracture energies per unit area from these spin-orbit calculations are presented in Table 1. Therefore, for simplicity, we present mostly the nonmagnetic calculations for the surface studies below.

Table 1. Calculated fracture energies per unit area without and with spin-orbit (s-o) interactions required to cleave along the three possible distinct layers in the crystal.

System	F (mJ/m^2)	F (s-o) (mJ/m^2)	M (missing bonds)
Cut1	2499	2186	4 U-Sb bonds at 0.325 nm
Cut2	4953	4441	5 U-Sb bonds (1 at 0.310, 4 at 0.311 nm)
Cut3	1179	1186	2 U-Sb bonds at 0.310 nm

For the (001) surface of USb₂, there are several possible surface terminations when the crystal is cleaved, as illustrated in Figure 2, where we show there are 5 possible cuts along the (001) plane at various z positions that will yield surfaces with different chemistries (Table 1).

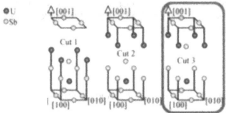

Fig. 2

There are 5 possible cuts: cut1 is between layer 1 and 2; cut2 is between layer 2 and 3; cut3 is between layer 3 and 4; cut4 is between layer 4 and 5; and cut5 is between layer 5 and 6. Because there is mirror symmetry along the z direction, cut5 and cut1 are the same, and cut4 and cut2 are the same. There are only 3 distinct cuts with regard to surface chemistry and structure. These cuts are cut1 (with a layer spacing of 0.244 nm between Sb and U layers), cut2 (with a layer spacing of 0.070 nm between of U and Sb) and cut3 (with a layer spacing of 0.244 nm between U and Sb).

We have calculated the fracture energy per unit area (F), in mJ/m², and missing bonds per area (M), (per unit surface area of 0.4274 nm x0.4274 nm) of these cuts and tabulated them in Table 1. The fracture energies per unit area for these three cuts are quite different. The lowest energy cut is cut3 at 1179 (or 1186 in spin-orbit calculations) mJ/m², while the other two cuts are about 2 and 4 times larger in energy. These drastic differences in fracture energies per unit area are much larger than observed in the cases of BaTiO₃, SrTiO₃ and CaTiO₃ perovskites or for metals [21-23]. If the fracture process on (001) USb₂ is determined solely by the fracture energy per unit area, then the most likely cleaving surface will be the cut3 surface, with the lowest fracture energy density of 1179 (or 1186 in spin-orbit calculations) mJ/m². The cut3 process will create two identical surfaces for the two parts that will be created. All these surfaces will terminate with a pure Sb layer on the surface, an underlying second layer of U atoms 0.070 nm below it, and a 3rd layer of Sb atoms separated from the U layer by 0.244 nm. This is followed by another layer of U atoms separated from the third layer by 0.244 nm.

With cut3 termination, we allowed the atoms of the top three layers to relax to the lowest energy with zero forces. The vertical relaxation was significant. The 1st -2nd layer spacing changed from 0.070 nm to 0.081 nm. This represents an outward expansion of the first layer of +15.7%, which is much larger than the ~1% change observed in fcc (111) or hcp (0001) surfaces [22]. It would be very informative to do surface structure experiments by LEED or ion scattering experiments to pin down the exact position of the atoms, which will help us understand how U and Sb atoms interact on the surface of USb₂. The 2nd -3rd layer spacing contracts from 0.244 nm to 0.238 nm (-2.5% contraction), which is the opposite of the 1st layer expansion, as one would expect from the charge compensation effects of next layers. The 3rd -4th layer spacing goes from 0.244 nm to 0.243 nm (-0.4% contraction) as the oscillation of these relaxations decays into the bulk [23].

In addition to the structural studies, we also performed studies of the charge transfer between atoms to explore the electronic interactions and properties in this compound and to help interpret the STM images. We calculated the LDOS on each atom and sorted it into contributions from different orbitals. We found most of the signal is from the $5f$ electrons of U, even though only about ~3 $5f$ electrons are present in a U atom. The much smaller contribution from Sb comes mainly from its d orbitals.

DISCUSSION

A comparison of the lattice parameter along the a-axis from the STM data to the published bulk value (and with the calculations) differs from the expected value by about 3%. The measured vertical layer spacing between the top surface and the layer below (0.080 nm) is very close to the value obtained from the calculated relaxed vertical layer spacing (0.081 nm), assuming cut3. Here the top surface is predicted to consist of Sb atoms with the second layer consisting of U atoms, which are offset from the Sb atoms by $a/2$ rather than resting directly below the missing atom positions. Although this situation makes interpretation of the STM vertical dimension problematic, we can expect the measured vertical dimension to be of the right order of magnitude.

The bond breaking energy calculations above point to a top surface consisting only of Sb atoms, but the charge calculations suggest that U atoms primarily contribute to the STM LDOS filled state maps, as much as over 29 to 1 for the relaxed surface. In effect, the STM sees almost exclusively the U atoms. Since the closed-loop filled and empty state images are nearly identical in structure intensity and there are no obvious systematic displacement of features, it appears that the STM sees the same atoms in both cases. This observation supports the interpretation that the dark features in the images are atomic vacancies. This interpretation encounters one difficulty with the energy argument, because the above process will require the surfaces with cut3, which have the lowest fracture energy density, to branch into cut2 terminations, which have the highest fracture energy per unit area. This particular cut is less energetically favorable, yet there are many black spots observed in our experiments (Figs. 1). One plausible explanation for the significant number of missing atoms from surfaces is that these vacancies were formed during crystal growth and that the reduced number of atoms and bonds weakens the bonding between those particular defected layers and is not associated with the higher energy cut2 fractures.

In conclusion, we have achieved what we believe to be the first atomic resolution STM images for a uranium compound taken at room temperature. The a, b, and c lattice parameters in the images confirm that the USb_2 crystals cleave on the (001) basal plane, as expected from our calculations. Further, our calculations indicate a symmetric cut between Sb planes to be the most favorable cleavage plane and that U atoms are responsible for the majority of the DOS signal measured by STM. A full report have been presented elsewhere [24].

We like to thank US DOE and NNSA for the support of this work at Los Alamos National Laboratory and at Stanford (SLAC contract DE-AC02-76SF00515).

REFERENCES

1. S. Lebegue, P.M. Oppeneer, and O. Eriksson, *Phys. Rev. B* **73**, (2006) p. 045119.
2. M.R. Castell, C. Muggelberg, and G.A.D. Briggs, *J. Vac. Sci. Technol. B* **14(2)**, (1995) p. 966.
3. M.R. Castell, C. Muggelberg, S.L. Dudarev, A.P. Sutton, G.A.D. Briggs, and D.T. Goddard, *Appl. Phys. A* **66**, (1998) p. S963
4. C. Muggelberg, M.R. Castell, G.A.D. Briggs, and D.T. Goddard, *Surf. Sci.* **404**, (1998) p. 673.
5. H. Norenberg and G. A. D. Briggs, Surf. Sci. 433-435 (1999) p. 127; U. Berner and K. Schierbaum, Thin Solid Films, 400 (2001) p. 46.
6. Z. Henkie, R. Maslanka, P. Wisniewski, R. Fabrowski, P.J. Markowski, J.J.M. Franse, and M. van Sprang, *J. Alloys and Compounds* **181**, (1992) p. 276
7. Z. Henkie, P. Wisniewski, R. Fabrowski, and R. Maslanka, *Solid State Comm.* **79(12)**, (1991) p. 1025
8. E. Guziewicz, T. Durakiewicz, C.G. Olson, J.J. Joyce, M.T. Butterfield, A.J. Arko, J.L. Sarrao, and A. Wojakowski, *Surf. Sci.* **600**, (2006) p. 1632
9. E. Guziewicz, T. Durakiewicz, M.T. Butterfield, C.G. Olson, J.J. Joyce, A.J. Arko, J.L. Sarrao, A. Wojakowski, and T. Cichorek, *Mat. Res. Soc. Symp. Proc.* **802**, (2004) p. 183
10. E. Guziewicz, T. Durakiewicz, M.T. Butterfield, C.G. Olson, J.J. Joyce, A.J. Arko, J.L. Sarrao, D.P. Moore, and L. Morales, *Phys. Rev. B* **69**, (2004) p. 045102
11. D. Aoki, P. Wisniewski, K. Miyake, R. Settai, Y. Inada, K. Sugiyama, E. Yamamoto, Y. Haga, and Y. Onuki, *Physica B* **281-282**, (2000) p. 71.
12. Y. Onuki, R. Settai, K. Sugiyama, Y. Inada, T. Takeuchi, Y. Haga, E. Yamamoto, H. Harima, and H. Yamagami, *J. Phys.: Condens. Matter* **19**, (2007) p. 125203
13. D. Aoki, P. Wisniewski, K. Miyake, N. Watanabe, Y. Inada, R. Settai, E. Yamamoto, Y. Haga, and Y. Onuki, *J. Phys. Soc. Japan* **68(7)**, (1999) p. 2182.
14. J. Leciejewicz, R. Troc, A. Murasik, and A. Zygmunt, *Phys. Stat. Sol.* **22**, (1967) p. 517
15. H. Kato, H. Sakai, K. Ikushima, S. Kambe, Y. Tokunaga, D. Aoki, Y. Haga, Y. Onuki, H. Yasuoka, and R.E. Walstedt, *Physica B* **359-361**, (2005) p. 1012
16. S. Tsutsui, M. Nakada, S. Nasu, Y. Haga, D. Aoki, P. Wisniewski, and Y. Onuki, *Phys. Rev. B* **69**, (2004) p. 054404
17. Z. Fisk and J. P. Remeika, "Growth of single crystals from molten metal fluxes" in *Handbook on the Physics and Chemistry of Rare Earths,* Vol. 12, edited by K. A. Gschneidner, Jr. and L. Eyring ~Elsevier, Amsterdam, (1989), p. 53; P. C. Canfield and Z. Fisk, Philos. Mag. B **65**, (1992) p. 1117
18. G. Kresse and J. Hafner, Phys. Rev. B **48**, 13 (1993) p. 115; G. Kresse and J. Furthmu"ller, Comput. Mater. Sci. **6**, (1996) p. 15.
19. P. E. Blo"chl, Phys. Rev. B **50**, (1994) p. 17953; G. Kresse and D. Joubert, Phys. Rev. B **59**, (1999) p. 1758.
20. A. M. N. Niklasson et al., Phys. Rev. B 67 (2003) p. 235105.
21. S. P. Chen, J. Mat. Res. 13, (1998) p. 1848.
22. S. P. Chen, Surface Science Lett. 264, (1992) p. L162.
23. S. P. Chen, A. F. Voter, and D. J. Srolovitz, Phys. Rev. Lett. 57, (1986) p. 1308.
24. S. P. Chen, M. Hawley, P. B. Van Stockum, et al., Phil. Mag. In press (2009).

Mater. Res. Soc. Symp. Proc. Vol. 1184 © 2009 Materials Research Society   1184-HH06-05

Size, Stability and Chemistry of Nanomaterials and Their Precursors by Mass Spectrometry Techniques

Jean-Jacques Gaumet[1], Didier Arl[1,2], Stéphane Dalmasso[2], Frédéric Aubriet[1] and Jean-Pierre Laurenti[2]

[1] LSMCL, Institut Jean Barriol, Université Paul Verlaine – Metz, ICPM, 1 bd Arago, 57070 Metz - France
[2] LPMD, Institut Jean Barriol, Université Paul Verlaine – Metz, ICPM, 1 bd Arago, 57070 Metz - France

ABSTRACT

Soft ionization mass spectrometry (MS) methods [Electro-Spray Ionisation - Fourier Transform Ion Cyclotronic Resonance MS (ESI-FTICRMS) and Matrix Assisted Laser Desorption Ionization coupled with Time of Flight MS (MALDI-TOFMS)] and associated fragmentation techniques appear to be an alternative way providing data on the size, stability and exact chemical composition of nanoparticles and their precursors, and potentially on interactions between particles. We report the application of both mass spectrometry techniques to analyze II-VI semiconductor nanomaterials (CdX with X = S or Se) and their organometallic precursors.

INTRODUCTION

Due to their size-dependent nature, nanomaterials characterization is a crucial issue as the size, shape, and dispersity must be accurately known for applications in device technology. For example, the modification of the absorption properties in respect of the size of II/VI nanoparticles is well known [1-2]. In particular, the maximum of absorption in UV-visible wavelength range shifts to UV when the radius of CdSe particles decreases; the color consequently varies from red to yellow [3]. To control the quality of the II-VI nanoparticles – also named nanocrystals (NCs) – their synthesis properties have to be improved to yield high quality NCs. Typically, size and size dispersity in these materials are measured by TEM imaging or estimated from the optical properties. Other physical and physico-chemical methods, such as NMR, X-ray diffraction, photoelectron spectroscopy and Raman spectroscopies are excellent tools for average analysis of a cluster or a NC [4-5]. However, specific distributions in the composition and structure of individual nanomaterials are not addressed. It is also crucial to get information from the precursors of these NCs, as it may help to explain the growing mechanism of nanomaterials. The use of mass spectrometry (MS) is an alternative way for getting some information about composition, size; surface, and stability when analyzing nanoparticles that are often generated by activation of organometallic precursors [6-8]. In this context we present the potential of soft ionization mass spectrometry techniques as tools for exploring II-VI nanomaterials and their precursors. We studied a full range of precursors and nanomaterials containing cadmium atoms surrounded either by sulphur or selenium atomic species. Electrospray ionization-Fourier transform ion cyclotron resonance mass spectrometry (ESI-FTICRMS) was used to analyze metal thiophenolate complexes used as precursors of semiconductors NCs. Additional structural and stability information have also been given by MS/MS dissociation performed on pseudo molecular ion. Matrix-assisted laser desorption ionization coupled to a time of flight MS (MALDI-TOFMS) allows to determine accurately the mass and the mass distribution of a series of CdSe NCs within the range 2-5 nm size. These results are consistent with those obtained by TEM and optical information for samples prepared by thermal growing process.

EXPERIMENTAL

Different cadmium thiophenolates were prepared by literature methods [9]. They correspond to the general formulae: $[Cd_{10}L_4(SPh)_{16}][(CH_3)_4N]_4$, named as "$Cd_{10}L_4$" and $[Cd_{17}L_4(SPh)_{28}][(CH_3)_4N]_2$, named as "$Cd_{17}L_4$" with L = S or Se. The SPh moiety is relevant of a thiophenolate, i.e. SC_6H_5 ligand. Figure 1 shows the general structure of both clusters. Note that each capped chalcogenide bound to a phenyl group not shown for clarity. CdSe NCs were prepared by thermal growing of $Cd_{10}Se_4$ in hexadecylamine (HDA, m.p. 47°C). HDA is used as a solvent and to functionalize NCs. The temperature was slowly increased (2°C.min⁻¹) and nanocrystal growth was monitored by periodic removal of small quantities and measurements of their absorbance spectra.

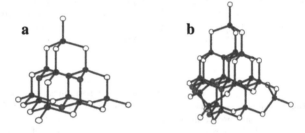

Figure 1. Structures of a) "$Cd_{10}L_4$" $[Cd_{10}L_4(SPh)_{16}][(CH_3)_4N]_4$, and b) "$Cd_{17}L_4$" $[Cd_{17}L_4(SPh)_{28}][(CH_3)_4N]_2$, black atoms are Cd, white atoms are the chalcogenes atoms (either S or Se), and red atoms are chalcogenes surrounded only by metal atoms.

ESI-FTICRMS analyses on the precursors were performed in negative ion mode by using an IonSpec HiRes FTICRMS (Varian-IonSpec Inc.) fitted with a 9.4 T actively shielded superconducting magnet and a Micromass Z-spray electrospray source. Freshly prepared samples ($\sim 10^{-5}$ mol.l⁻¹ in dry acetonitrile) were directly infused in the source at a flow rate of 3 ml.mn⁻¹. The high voltage varied from -3800 to -4200 V. Source and probe temperatures were fixed at 90 and 100°C, respectively, to avoid thermal degradation of organometallic samples during ESI step. Nitrogen was used as drying and nebulizing gas. Sample cone voltage was set from -15 V to -130 V and the extraction cone voltage was kept constant at – 10 V. Confirmation of the assigned ionic species was made by both exact mass measurement and by comparing the experimental data with theoretical isotopic pattern calculated by Omega 8 Exact Mass Calculator software (Varian-IonSpec Inc). MS/MS experiments were conducted after parent ion isolation. Amongst the different activation modes, the Sustained Off Resonance Irradiation – Collision Induced Dissociation (SORI-CID) method was used with a duration of the SORI excitation of 250 ms, an excitation amplitude set at 9 V and a collision gas (N_2) pulse duration kept at 10 ms.

MS analyzes on thermally grown NCs were performed using a MALDI-TOFMS Bruker Reflex IV with dithranol as a matrix. Desorption and ionization of the samples were achieved by irradiation with a pulsed nitrogen laser (337 nm, E = 3.68 eV, pulse duration ~ 4 ns , maximum pulse energy ~ 300 μJ). To increase the ionization yield, CdSe NCs were functionalized with 2-aminoethanethiol (AET, 99% purity). After dilution in DMF of the CdSe NCs functionalized with HAD, an excess of AET is added and the solution is heated at 70°C for 5 min. The addition of methanol to the NCs solution is followed by centrifugation. TEM was performed on a Philips CM 200 operating at 200 kV in the bright field mode. Size and size distributions were obtained by collecting manually 100 values for each images saved from the digital micrograph. For absorption measurements, a white light produced by a Quartz-Iode source illuminates the NCs diluted in toluene at room temperature. The transmitted light is dispersed by grating and analyzed by CCD camera with a resolution of 0.23 nm.

RESULTS AND DISCUSSION

Specific fingerprint and stability of [Cd$_{17}$S$_4$(SPh)$_{28}$][(CH$_3$)$_4$N]$_2$

Electrospray coupled with FTICRMS allows the rapid analysis of nanomaterial precursors. Even if measurements were performed in both ion modes, more relevant information has been founded in the negative ion mode. Figure 2 shows single ESI-FTICRMS mass spectrum in the negative ion mode for Cd$_{17}$S$_4$ cluster at -30 V cone voltage.

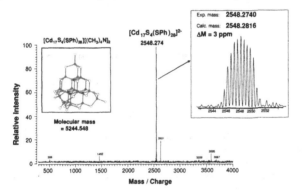

Figure 2. ESI-FTICR mass spectrum of [Cd$_{17}$S$_4$(SPh)$_{28}$][(CH$_3$)$_4$N]$_2$ at a cone voltage of -30 V in negative ion mode.

A strong peak at m/z 2548.27 corresponds to the [Cd$_{17}$S$_4$(SPh)$_{28}$]$^{2-}$ intact ion, with the removal of the two [(CH$_3$)$_4$N]$^+$ counterions. Note that the increase of the cone voltage induces the detection of ions including 10-16 cadmium atoms and several ions detected in a previous study related to Cd$_{10}$L$_4$ compound like [Cd(SPh)$_3$]$^-$ and (SPh)$^-$ at m/z 440.93 and 109.01 respectively [10].

ESI-MS/MS analysis of [Cd$_{17}$S$_4$(SPh)$_{28}$]$^{2-}$ species provides some explanations on the stability of "Cd$_{17}$L$_4$" and also on all metal thiophenolates that are potential precursors of semiconductor nanomaterials. Figure 3 shows the fragmentation of this ion by SORI-CID.

[Cd(SPh)₃]⁻ and (SPh)⁻ at m/z 440.93 and 109.01 respectively, are the most intense ions detected. They are relevant to the loss of the capping faces and tetrahedral corners of the cluster. However 4 sulfur atoms are only bounded with cadmium atoms. The Cd_4S_4 moiety in the $[Cd_4S_4[Cd(SPh)_2]_n(SPh)]^-$ ions with n = 2 to 5 (m/z between 1250 and 2500) detected after MS/MS is clearly associated to these 4 specific sulfur atoms. Four detected daughter ions are indicative of either $Cd(SPh)_2$ loss (mass difference : 330.4) or oligomeric fragments of same nature. No $[Cd_x(SPh)_{2x+1}]^-$ ion species were founded in the mass spectra except the "surrounding" signature with x = 1, $[Cd(SPh)_3]^-$. This fact illustrates the remarkable stability of the "Cd_4L_4" cluster core. The lightest detected ion species was $[Cd_4S_4[Cd(SPh)_2]_2(SPh)]^-$ at m/z 1348. This structurally corresponds to a Cd_4S_4 surrounded by $Cd(SPh)_2$ ligands. Same behavior was found previously in similar work with "$Cd_{10}L_4$" thiophenolate precursor series. Some theoretical calculations are under progress at that level for determining the most stable and probable structure (e.g. a cubic form for this elementary piece) [10].

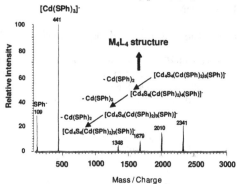

Figure 3. SORI-CID ESI-FTICRMS/MS of $[Cd_{17}S_4(SPh)_{28}]^{2-}$ at m/z 2548 in negative ion mode.

Size and dispersity information by MALDI-TOF mass spectrometry

Figure 4 shows MALDI-TOF positive ion mode mass spectra for some functionalized CdSe-AET NCs that were grown thermally in hexadecylamine at 196°C (4a) and 229°C (4b) respectively. Their size determined by TEM measurements show an average diameter of 2.3 and 2.7 nm respectively. Figure 5 recapitulates the data for the 2.7 nm CdSe NCs including size distribution (+/- 0.36 nm). Typical Gaussian-shaped peak centered at m/z values of roughly 22 000 (4a) and 40 000 (4b) are observed in the mass spectra with the general formula $[M_{NC}]^+$. Observation of a second peak for the smallest NCs is the consequence of the laser power used for getting sample desorption. When running the MALDI-TOF MS experiments, laser power was tuned to 90 % for the 2.2 nm compared to 55 % for the 2.7 nm NCs to favor the detection of the highest mass peak. The second peak at m/z 11 000 is assignable to a doubly charged nanocrystal species $[M_{NC}]^{2+}$. This experiment suggests that the MALDI process only results in charge addition and not in fragmentation of the parent ion. Even if the low resolution observed in linear mode for our experiments did not allow the fwhm and size dispersion data on CdSe NCs samples, these results concerning CdSe for singly and doubly charged species are consistent with a significant absorption cross section of CdSe at 337 nm laser [11].

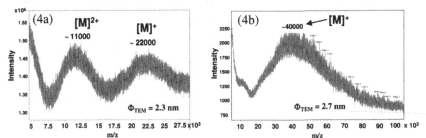

Figure 4. MALDI-TOF mass spectrum of 2 size thermal grown NCs of CdSe functionalized by AET and using dithranol as matrix. Growth temperatures were 196°C (2.3 nm) (4a) and 229°C (2.7 nm) (4b), respectively.

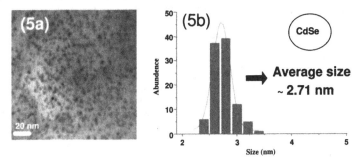

Figure 5. TEM image of CdSe NCs deposited on a 200-mesh Cu grid with 480 000 magnification (5a) and the size distribution obtained on the same sample (5b).

However, the quantum dot size can be estimated from the number of repeating CdSe moieties $(CdSe)_n$ in the NC, by using Inoue model [12].

$$(CdSe)_n = (\pi/2)N_A(d^3)/(3V_m) \qquad (1)$$

N_A is Avogadro's number, d is the NC diameter and V_m is the molar volume of bulk CdSe ($V_m = 3.35 \times 10^{22}$ nm^3.mol^{-1}). Knowing $(CdSe)_n$ allows to deduce from eq. 1 the size of the NCs. From the software Exact Mass Calculator, (Ionspec Inc.), it is possible to estimate the number of CdSe units. For $M_{NC}^+ = 40\ 000$, n = 222 CdSe units, and for $M_{NC}^+ = 22\ 000$, n = 122 CdSe units. Consequently the average diameter of the NCs was calculated for both samples. We obtained 2.78 nm and 2.28 nm, respectively, which are consistent when considering the size distribution for the two CdSe nanocrystals analyzed by TEM (see table 1). All these results show the versatility of soft ionization MS techniques and their complementarity to other analytical tools necessary for determining the size and the composition of nanomaterials.

Table 1. Comparison of NCs diameters between TEM image and MALDI-TOFMS calculations.

TEM diameter (nm)	Mass determined by MALDI-TOFMS	n : number of CdSe units calc.	Average diameter from MALDI-TOFMS data (nm)	Absorption wavelength (nm)
2.20 +/- 0.35	22 000	122	2.28	472.2
2.70 +/- 0.36	40 000	222	2.78	490.7

CONCLUSIONS

This study shows all the potential of mass spectrometry for analyzing II-VI nanomaterials and their precursors. The quality of the precursor's synthesis can be easily checked, and ESI FTICRMS provides specific fingerprint, structural and stability information for all precursors. The MALDI-TOF MS is a powerful tool for analyzing size and size distribution of inorganic nanomaterials and is complementary to other spectroscopy and electronic microscopy techniques. Further optimization of experimental conditions for increasing mass resolution and starting to analyze III-V nanosemiconductors materials are in progress in our group.

ACKNOWLEDGMENTS

The authors would like to thank Pr. G. Strouse for helpful discussions concerning nanomaterials growing and Pr. G. Kirsch for the help in the synthesis of nanomaterials and their precursors. D. Arl acknowledges the Ministère de l'Enseignement Supérieur et de la Recherche (France).

REFERENCES

1. T. Vossmeyer, L. Katsikas, M.Giersig and I.G. Popovic, *J. Phys. Chem.* **9**, 7665 (1994).
2. S.L. Cumberland, K.M. Hanif, A. Javier, G. Khitrov, G.F. Strouse, S.M. Woessner and C.S. Yun, *Chem. Mater.* **14**, 1576 (2002).
3. A.P. Alivisatos, *J. Phys. Chem. B* **100**, 13226 (1996).
4. M.B. Mohamed, D. Tonti, A. Al-Salman, A. Chemseddine and M. Chergui, *J. Phys. Chem. B* **109**, 10533 (2005).
5. B.F.G. Johnson and J.S. Mc Indae, *Coordination Chemistry Reviews* **115**, 8706 (2000).
6. T. Löver, W. Henderson, G.A. Bowmaker, J.M. Seakins and R.P. Cooney, *Inorg. Chem.* **36**, 3711 (1997).
7. J.J. Gaumet, G.A. Khitrov and G.F. Strouse, *Nanoletters* **2**, 375 (2002).
8. J.J. Gaumet, G.A. Khitrov and G.F. Strouse, *Mat. Sci. Engineering C*, **19**, 299 (2002).
9. I.G. Dance, A. Choy and M.L. Scudder, *J. Am. Chem. Soc.*, **106**, 6285 (1984).
10. D. Arl, F. Aubriet and J.J. Gaumet, *J. Mass Spectrom.* in press (2009).
11. G.A. Khitrov and G.F. Strouse, *J. Am. Chem. Soc.*, **125**, 10465 (2003).
12. H. Inoue, N. Ichiroku, T. Torimoto, T. Sakata, H. Mori and H. Yoneyama, *Langmuir*, **10**, 4517 (1994).

Mater. Res. Soc. Symp. Proc. Vol. 1184 © 2009 Materials Research Society 1184-HH07-01

Characterization of Nanostructured Organic-Inorganic Hybrid Materials Using Advanced Solid-State NMR Spectroscopy

Kanmi Mao[1,2], Jennifer L. Rapp[1], Jerzy W. Wiench[1] and Marek Pruski[1,2]
[1]U.S. DOE Ames Laboratory, Iowa State University, Ames, IA 50011, USA
[2]Department of Chemistry, Iowa State University, Ames, IA 50011, USA

ABSTRACT

We demonstrate the applications of several novel techniques in solid-state nuclear magnetic resonance spectroscopy (SSNMR) to the structural studies of mesoporous organic-inorganic hybrid catalytic materials. Most of these latest capabilities of solid-state NMR were made possible by combining fast magic angle spinning (at $\geq$ 40 kHz) with new multiple RF pulse sequences. Remarkable gains in sensitivity have been achieved in heteronuclear correlation (HETCOR) spectroscopy through the detection of high-γ (^{1}H) rather than low-γ (e.g., ^{13}C, ^{15}N) nuclei. This so-called indirect detection technique can yield *through-space* 2D ^{13}C-^{1}H HETCOR spectra of surface species under natural abundance within minutes, a result that earlier has been out of reach. The ^{15}N-^{1}H correlation spectra of species bound to a surface can now be acquired, also without isotope enrichment. The first indirectly detected *through-bond* 2D ^{13}C-^{1}H spectra of solid samples are shown, as well. In the case of 1D and 2D ^{29}Si NMR, the possibility of generating multiple Carr-Purcell-Meiboom-Gill (CPMG) echoes during data acquisition offered time savings by a factor of ten to one hundred. Examples of the studied materials involve mesoporous silica and mixed oxide nanoparticles functionalized with various types of organic groups, where solid-state NMR provides the definitive characterization.

INTRODUCTION

Solid-state nuclear magnetic resonance (SSNMR) spectroscopy is becoming a very valuable technique for the characterization of nanomaterials, as the quest for achieving solution-like resolution has been further advanced by the development of fast magic angle spinning (MAS) at rates of up to 70 kHz [1-3]. Advantages of fast MAS include the possibility of using low-power radiofrequency (rf) decoupling schemes [4-6], minimization/elimination of spinning sidebands and increased frequency range in the indirect dimension of rotor-synchronized experiments. In many applications, the greatest advantage is the reduction, and in some cases practical elimination, of homonuclear dipolar couplings between high-γ nuclei, such as ^{1}H and ^{19}F. The ability to decouple ^{1}H nuclei from each other by means of fast MAS has led to the development of 2D HETCOR experiments, which rely on the detection of high-γ (sensitive) ^{1}H nuclei instead of the low-γ (insensitive) nuclei, such as ^{13}C [7-13]. This technique, referred to as indirect detection, offers considerable enhancements in sensitivity, which can reduce the acquisition time by a factor of more than 10, in some cases [10]. Typically, the internuclear exchange of magnetization in HETCOR experiments occurs via cross polarization (CP), i.e., relies on dipolar interaction between spins, thereby providing *through-space* correlations. Most recently, the utilization of much weaker scalar (J) couplings has become practical, providing *through-bond* correlations in a manner similar to solution-state INEPT NMR experiments [14,15]. Since the key challenge of extending the lifetime of ^{1}H and ^{13}C transverse relaxation during the INEPT transfer can be met by using fast MAS and efficient ^{1}H-^{1}H decoupling [14], we recently introduced the indirectly detected *through-bond* HETCOR experiment in the solid-state [15].

Another valuable technique that takes advantage of the efficient homonuclear decoupling provided by fast MAS in conjunction with low power heteronuclear decoupling capabilities, uses the Carr-Purcell-Meiboom-Gill (CPMG) refocusing [16] to enhance the sensitivity in 1D and 2D ^{29}Si experiments [17-19]. This increase in sensitivity is made possible by extending the transverse relaxation time T_2 of silicon under fast MAS and ^{1}H heteronuclear decoupling, which permits multiple refocusing of transverse magnetization by a series of rotor-synchronized π-pulses, thereby improving the time performance of ^{29}Si NMR by one to two orders of magnitude. The CPMG refocusing has enabled measurements of 2D ^{29}Si{^{1}H} HETCOR spectra of organic-inorganic hybrid materials [17-19]. Such experiments were difficult to carry out via conventional methods due to the lack of sensitivity.

As a result of these developments, the collection of highly resolved 2D HETCOR spectra in samples containing sub-milligram amounts of naturally abundant species has become possible [10,17]. We provide a short overview of these methods and demonstrate how they can be applied to characterize the structure of mesoporous organic-inorganic hybrid catalytic materials.

EXPERIMENT

The solid-state NMR experiments were performed at 9.4 and 14.1 T on a Chemagnetics Infinity spectrometer equipped with a 1.8 mm double-tuned probe (A. Samoson) and a Varian NMR System 600 spectrometer equipped with a 1.6 mm triple resonance FastMAS® probe, respectively. The experimental parameters are included in the figure captions, where B_0 denotes the static magnetic field, ν_R magic angle spinning rate, ν_{RF}^H and ν_{RF}^X the magnitudes of radiofrequency magnetic fields applied to ^{1}H and X spins, τ_{CP} the cross polarization time, τ_{RD} the recycle delay, NS the number of scans, N_{CPMG} the number of echoes acquired in CPMG experiments, and τ_{CPMG} the time interval between π pulses. All chemical shifts are reported on the δ scale and referenced to TMS for ^{1}H, ^{13}C, ^{29}Si and neat nitromethane (CH_3NO_2) for ^{15}N, all at 0 ppm.

DISCUSSION

Studies of surface-bound molecules

The effect of fast magic angle spinning (MAS) on the spectral line shape is demonstrated in 1D ^{1}H MAS spectra of allyl groups (-CH_2-CH=CH_2) covalently bound to the surface of mesoporous silica nanoparticles (MSNs) (figure 1). The improvement in spectral resolution with increasing spinning rate is quite clear. Although the ^{1}H-^{1}H homonuclear dipolar coupling has not been completely removed by spinning the sample at 40 kHz, the resolution is sufficient to allow for unambiguous structural assignments of ^{1}H resonances.

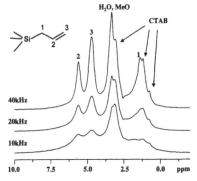

Figure 1. ^{1}H MAS NMR spectra of allyl groups covalently bound to the surface of MSNs, acquired under the following conditions: $B_0 = 9.4$ T, $v_R = 10$, 20 and 40 kHz, $v_{RF}^H = 170$ kHz, $\tau_{RD} = 0.4$ s, NS = 64 [17].

The line narrowing by fast MAS enabled the acquisition of indirectly detected 2D HETCOR spectra, such as the one shown in figure 2. Depicted in this figure is the ^{13}C–^{1}H indirectly detected HETCOR spectrum of MSNs functionalized with rhodium-phosphine complexes, the structure of which is shown at the bottom of the spectrum. The ^{13}C-^{1}H cross-peaks can be easily assigned to various components of the phosphine complexes, as indicated in the figure.

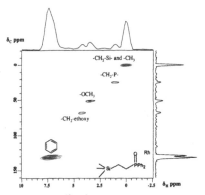

Figure 2. 2D ^{13}C–^{1}H indirectly detected HETCOR spectrum of MSNs functionalized with rhodium phosphine complexes, measured under the following conditions [20]: $B_0 = 14.1$ T, v_R

= 35 kHz, v_{RF}^H during excitation = 62.5 kHz, v_{RF}^H during CP = 54 kHz, v_{RF}^C during CP = 81 kHz, v_{RF}^C during SPINAL-64 decoupling = 9 kHz, τ_{RD} = 1 s, τ_{CP} = 1 ms, NS = 32, number of rows = 160, t_1 increments = 25 μs. Following our earlier report [10], a single π pulse with v_{RF}^H = 62.5 kHz was used for ^{1}H decoupling during the evolution period.

The evidence that the rhodium-phosphine complexes are surface-bound has been provided by the ^{29}Si NMR spectra (not shown), which exhibited the resonances characteristic of Si-R functionalities (the so-called T sites). By integrating the ^{29}Si spectra, we found that approximately 3.8% of the silicon sites are functionalized with the complexes.

The time savings achieved through the indirect detection can be remarkable. The spectrum in Figure 2 was collected in 2.5 hours, whereas it would have taken ~18 hours to obtain a similar result using the conventional HETCOR methods. Another noteworthy example of the sensitivity enhancement is the indirectly detected ^{15}N-^{1}H HETCOR spectrum of nitrogen containing ethylureidophenyl (EUP) groups, which are also covalently attached to the surface of mesoporous silica (figure 3). Although it took 48 hours to complete this measurement, it represents, to the best of our knowledge, the first demonstration of ^{15}N-^{1}H correlations in naturally abundant species in solids. The studied sample contained only ~2 μmols of the organic functional groups.

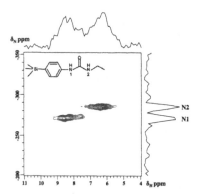

Figure 3. 2D ^{15}N-^{1}H indirectly detected HETCOR spectrum of EUP-MSN silica, acquired under the following conditions: B_0 = 14.1 T, v_R = 40 kHz, v_{RF}^H during excitation and heteronuclear decoupling via a single π pulse [10] = 100 kHz, v_{RF}^H during CP = 60 kHz, v_{RF}^N during CP = 100 kHz, v_{RF}^N during SPINAL-64 decoupling = 11 kHz, τ_{CP} = 1 ms, τ_{RD} = 1 s, number of rows = 75, t_1 increment = 25 μs, NS = 1024.

The measurement of 2D spectra of species bound to a surface via indirect detection has been recently extended to include *through-bond* correlations using the refocused INEPT (INEPTR) method for polarization transfer [15]. Fast MAS proved instrumental in achieving high efficiency in this experiment, because it helped to prevent the decoherence of transverse magnetization of both ^{1}H and ^{13}C nuclei. The indirectly detected 2D ^{13}C-^{1}H CP- and INEPT-HETCOR spectra of MSN functionalized with 3-(pentafluorophenyl)propyl groups (PFP-MSN) are compared in figure 4. The correlations representing long-range interactions (circled in red) in the indirectly detected CP-HETCOR spectrum (figure 4a) are eliminated in the INEPT-HETCOR spectrum (figure 4b), clearly exhibiting the efficacy of this technique for the measurement of *through-bond* connectivities in naturally abundant materials. The outstanding sensitivity offered by indirect detection provides basis for the development of other advanced 2D and 3D correlation methods.

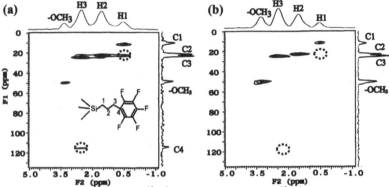

Figure 4. 2D indirectly detected ^{13}C-^{1}H CP-HETCOR (a) and ^{13}C-^{1}H INEPT HETCOR (b) spectra of PFP-MSN, acquired under the following conditions [15]: $B_0 = 14.1$ T, $\nu_R = 40$ kHz, ν_{RF}^H during short pulses = 110 kHz, ν_{RF}^H during tangent ramp CP = 60 kHz, ν_{RF}^C during short pulses and CP = 100 kHz, ν_{RF}^C during SPINAL-64 decoupling = 10 kHz, $\tau_{CP} = 4.5$ ms, $\tau_{RD} = 1$ s, NS = 48 (a) and 128 (b), number of rows = 160; t_1 increment = 25 µs, $\tau_1 = 0.6$ ms, $\tau_2 = 0.8$ ms, with acquisition times of 4.5 h (a) and 12.5 h (b).

Another means for improving the sensitivity, applicable to nuclei with long transverse relaxation times, is via incorporation of the Carr-Purcell-Meiboom-Gill (CPMG) refocusing during the acquisition period. Fast MAS can play an instrumental role in the application of this technique to mesoporous organic-inorganic hybrid materials, as it allows for more efficient refocusing of the magnetization of ^{29}Si nuclei that reside in the vicinity of hydrogen [17,18]. The gain in signal intensity achievable by this method is demonstrated in figure 5a, where the 1D ^{1}H-^{29}Si CPMAS spectrum measured with CPMG refocusing is compared with one obtained via the standard free induction decay experiment. Both spectra were acquired with the same experimental time, yet differ in the signal to noise ratio by a factor of approximately 10. The extension of this experiment to 2D spectroscopy is shown in figure 5b. The 2D ^{1}H-^{29}Si spectrum reveals the spatial proximity between allyl groups and the silica surface. The presence of the strong correlation between H3 protons and the Q^4 silica sites shows unambiguously that the functional groups reside on the surface in a prone position.

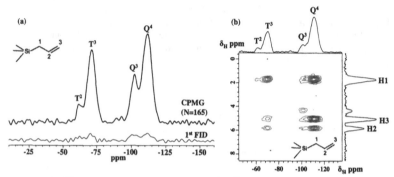

Figure 5. ^{1}H-^{29}Si CP vs. CP-CPMG (**a**) and 2D ^{1}H-^{29}Si CPMG HETCOR spectra (**b**) of allyl groups covalently bound to the surfaces of mesoporous silica nanoparticles, acquired under the following conditions [18]: $B_0 = 14.1$ T, $v_R = 40$ kHz; for (**a**): v_{RF}^H during excitation = 100 kHz, v_{RF}^H during tangent ramp CP = 60 kHz, v_{RF}^H during SPINAL-64 decoupling = 12 kHz, v_{RF}^{Si} during CP = 100 kHz, $\tau_{CP} = 8$ ms, $\tau_{RD} = 0.8$ s, NS = 320; for (**b**): same as (**a**), except with CPMG, $N_{CPMG} = 165$, and $\tau_{CPMG} = 6$ ms.

Studies of the bulk mesoporous structure

The CPMG technique can be also exploited in ^{27}Al-^{29}Si correlation spectroscopy of bulk materials. Acquisition of ^{27}Al-^{29}Si spectra is inherently challenging due to low efficiency of the CP process and low concentration of ^{27}Al-^{29}Si spin pairs, especially in materials with high Si/Al ratio. We have demonstrated that the sensitivity of ^{27}Al-^{29}Si correlation spectroscopy can be considerably increased by the combined use of RAPT (rotor assisted population transfer [21]) and CPMG data acquisition [22].

This technique provided first direct evidence that all Al atoms in mesoporous aluminum silicate catalysts (Al-MS) are surrounded by Si atoms forming Al-O-Si bonds in the framework. The ^{1}H-^{29}Si CPMAS, ^{29}Si DPMAS, and ^{27}Al-^{29}Si CPMAS spectra of Al-MS are shown in figure 6.

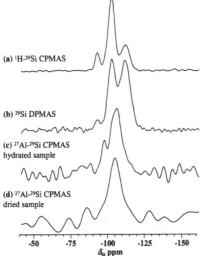

(a) ^{1}H-^{29}Si CPMAS

(b) ^{29}Si DPMAS

(c) ^{27}Al-^{29}Si CPMAS
hydrated sample

(d) ^{27}Al-^{29}Si CPMAS
dried sample

-50   -75   -100   -125   -150
δ_{Si} ppm

Figure 6. ^{1}H-^{29}Si CPMAS **(a)**, ^{29}Si DPMAS **(b)** and ^{27}Al-^{29}Si CPMAS **(c,d)** spectra of Al-MS catalyst with Si/Al ratio of around 60 [22]. The sample was studied in a hydrated state **(a)-(c)** and after drying at 100 °C **(d)**. All spectra were measured at $B_0 = 14.1$ T under 10 kHz MAS. Spectrum **(a)** was acquired using $v_{RF}^H = 96$ kHz during initial excitation, v_{RF}^H during CP = 34-40 kHz, $v_{RF}^{Si} = 34$ kHz during CP and CPMG, $\tau_{CP} = 8$ ms, $\tau_{CPMG} = 4$ ms, N$_{CPMG} = 6$, $\tau_{RD} = 1$ s and NS = 3000. For direct excitation of ^{29}Si **(b)**, the following parameters were used: $\tau_{RD} = 300$ s and NS = 300. Spectra **(c,d)** were obtained using the RAPT-CP-CPMG sequence with $v_{RF}^{Al} = 6$ kHz during RAPT, $\tau_{RAPT} = 1.2$ ms, $v_{RF}^{Al} = 1.5$ kHz during CP, v_{RF}^{Si} during CP = 7-8 kHz, $\tau_{CP} = 6$ ms, $v_{RF}^{Si} = 26$ kHz during CPMG, N$_{CPMG} = 6$, $\tau_{CPMG} = 4$ ms, $\tau_{RD} = 10$ ms (after the acquisition of the last echo), and NS = 6,000,000.

Deconvolution of the silicon resonances in the DPMAS spectrum (figure 6b) provided good estimates of relative percentages of Q^n(0Al) sites (n=2,3,4). However, due to the low aluminum content and signal overlap, ^{29}Si resonances corresponding to Q^n(mAl) (m >0) sites could not be distinguished in either CP or DPMAS spectra (figure 6a and b). On the other hand, utilizing the RAPT-CPMAS-CPMG scheme [22] allowed for selective observation of resonances representing Q^4(1Al) and Q^3(1Al) silicon sites, observed at -104 an -96 ppm, respectively (figure 6c). Only one silicon resonance, from the Q^4(1Al) sites at -104 ppm, is observed in figure 6d, which confirms the above assignments. The absence of the resonance from Q^3(1Al) sites is a result of their location near the pore surface. Upon drying, the quadrupole coupling constant of ^{27}Al nuclei associated with these sites increases and they are no longer able to efficiently polarize the neighboring ^{29}Si nuclei. It is important to note that the spectra of figure 6 were collected at a slower spinning speed (10 kHz); however, an overall sensitivity gain by a factor of 5 was still attainable. This Al-MS catalyst showed an excellent reactivity for the Claisen rearrangement of allyl phenyl ether, despite the low surface concentration of Al.

CONCLUSION

We have demonstrated several new capabilities of solid-state NMR spectroscopy, brought about by MAS at 40 kHz and/or improved pulse sequences, as well as their applications to the studies of mesoporous materials. The indirect detection of low-γ nuclei, which until recently has been impractical due to the lack of adequate ^{1}H homonuclear decoupling capabilities, enabled faster acquisition of ^{13}C-^{1}H and ^{15}N-^{1}H HETCOR spectra. It also became possible to utilize the J-couplings for polarization transfer to generate indirectly detected *through-bond* correlations between heteronuclei. The CPMG refocusing offered a possibility of enhancing the sensitivity in ^{29}Si NMR. These methods can now be applied to study the mesoporous nanoparticles in a routine manner. In particular, they can be used to (1) detail the structure of non-functionalized bulk materials and surfaces, (2) study the structure and absolute/relative concentration of various moieties inside the mesopores under the natural abundance, (3) determine their spatial distribution, orientation with respect to the surface and dynamic behavior, and (4) monitor the catalysts' stability under the reaction conditions.

ACKNOWLEDGMENTS

This research at the Ames Laboratory was supported by the U.S. Department of Energy, Office of Basic Energy Sciences, under Contract No. DE-AC02-07CH11358. The authors would like to thank Dr. Victor S. -Y. Lin and his group members, Hung-Ting Chen, Chih-Hsiang Tsai, Seong Huh, Yulin Huang and Kasey J. Strosahl for providing the MSN samples studied in this work.

REFERENCES

1. Ago Samoson, in *Encyclopedia of Nuclear Magnetic Resonance*, edited by D.M.R. Grant, K. Harris (John Wiley & Sons, Chichester, 2002), p. 59.
2. A. Samoson, T. Tuherm, Z. Gan, Solid State Nucl. Magn. Reson. 20, 130 (2001).
3. L. S. Du, A. Samoson, T. Tuherm, C. P. Grey, Chem. Mater. 12, 3611 (2000).
4. M. Kotecha, N. P. Wickramasinghe, Y. Ishii, Magn. Reson. Chem. 45, S221 (2007).
5. M. Ernst, A. Samoson, B. H. Meier, Chem. Phys. Lett. 348, 293 (2001).
6. M. Ernst, M. A. Meier, T. Tuherm, A. Samoson, B. H. Meier, J. Am. Chem. Soc. 126, 4764 (2004).
7. Y. Ishii and R.Tycho, J. Magn. Reson. 142, 199 (2000)
8. Y. Ishii, J. P. Yesionowski, R. Tycho, J. Am. Chem. Soc. 123, 2921 (2001)

9. F. K. Paulson, C. R. Morcombe, V. Gaponenko, B. Danchck, R. A. Byrd, K. W. Zilm, J. Am. Chem. Soc. 125, 15831 (2003).
10. J. W. Wiench, C. E. Bronnimann, V. S. -Y. Lin, M. Pruski, J. Am. Chem. Soc. 129, 12076 (2007).
11. B. Reif and R. G. Griffin, J. Magn. Reson. 160, 78 (2003).
12. D. H. Zhou, G. Shah, M. Cormos, C. Mullen, D. Sandoz, C. M. Rienstra, J. Am. Chem. Soc. 129, 11791 (2007).
13. D. H. Zhou and C. M. Rienstra, Angew. Chem. Int. Ed. 47, 7328 (2008).
14. B. Elena, A. Lesage, S. Steuernagel, A. Böckmann, L. Emsley, J. Am. Chem. Soc. 127, 17296 (2005).
15. K. Mao, J. W. Wiench, V. S. -Y. Lin, M. Pruski, J. Magn. Reson. 196, 92 (2009).
16. S. Meiboom and D. Gill, Rev. Sci. Instr. 29, 688 (1966).
17. J. Trebosc, J. W. Wiench, S. Huh, V. S. -Y. Lin, M. Pruski, J. Am. Chem. Soc. 127, 7587 (2005).
18. J. W. Wiench, V. S. -Y. Lin, M. Pruski, J. Magn. Reson. 193, 233 (2008).
19. J. W. Wiench, Y. S. Avadhut, N. Maity, S. Bhaduri, G. K. Lahiri, M. Pruski, S. Ganapathy, J. Phys. Chem. B, 111, 3877 (2007).
20. J. L. Rapp, Y. Huang, M. Natella, Y. Cai, V. S.-Y. Lin, M. Pruski, Solid State Nucl. Magn. Reson., 35, 82 (2009).
21. Z. Yao, H. -T. Kwak, D. Sakellariou, L. Emsley, P. J. Grandinetti, Chem. Phys. Lett., 327, 85 (2000).
22. Y. Cai, R. Kumar, W. Huang, B. G. Trewyn, J. W. Wiench, M. Pruski, V. S. -Y. Lin, J. Phys. Chem. C 111, 1480 (2007).

Mater. Res. Soc. Symp. Proc. Vol. 1184 © 2009 Materials Research Society   1184-HH08-04

Backside Analysis of Ultra-Thin Film Stacks in Microelectronics Technology Using X-ray Photoelectron Spectroscopy

Thomas Hantschel[1], Cindy Demeulemeester[1], Arnaud Suderie[1], Thomas Lacave[1], Thierry Conard[1], and Wilfried Vandervorst[1,2]

[1]IMEC, Kapeldreef 75, B-3001 Leuven, Belgium.

[2]Instituut voor Kern- en Stralingsfysica, K. U. Leuven, Celestijnenlaan 200D, B-3001 Leuven, Belgium.

ABSTRACT

X-ray photoelectron spectroscopy (XPS) has become increasingly important over the past few years for supporting the development of ultra-thin layers for high-k metal gates. As the analysis depth of XPS is however limited to about 5-7 nm, it would be extremely useful if the analysis could be carried out from the backside using standard silicon wafers. This approach puts extreme requirements on the sample preparation as hundreds of micrometers of bulk silicon have to be removed and one has to stop with nanometer precision when reaching the interface to the ultra-thin layer stack. Therefore, we have developed dedicated procedures for preparing and analyzing samples for backside XPS analysis. This paper presents the developed approach with a focus on sample preparation using plan-parallel polishing, endpoint detection by interference fringes, and selective wet etching. First angle-resolved XPS (ARXPS) analysis results of metal gate stacks demonstrate the power of such backside analysis.

INTRODUCTION

The introduction of new materials, the use of more complex materials systems and the down-scaling of layer thicknesses in microelectronics technology poses big challenges in the metrology area. The use of ultra-thin layers for high-k metal gates for improved transistor performance requires an analysis technique which is extremely surface sensitive and can provide information about the stack composition, interaction at the interfaces between the different layers and the influence of different processing steps on the stack composition. X-ray photoelectron spectroscopy (XPS) is gaining increasing importance for this task in the last few years. As the analysis depth of XPS is however limited to about 5-7 nm, and to avoid perturbation of the XPS signal from top surface contamination, the possibility to perform the analysis also from the wafer backside is urgently needed in order to gain access to the region of interest. This requires the removal of hundreds of micrometers thick bulk silicon and at the same time one has to stop at the layer of interest with nanometer-scale accuracy. To tackle this enormous challenge, we have developed dedicated sample preparation procedures and applied the backside XPS approach to advanced material systems. The key of the developed method is an optimized procedure for plan-parallel polishing (extension of tripod-based polishing [1]), the use of interference fringes for thickness measurement and the removal of silicon by selective wet etching. The basic approach for backside analysis has been developed originally for secondary ion mass spectrometry (SIMS) [2,3] but the requirements for backside XPS samples are even more stringent as the silicon substrate has to be removed completely for backside XPS analysis whereas a few hundreds of nanometers thick silicon layer can remain on the sample for backside SIMS analysis. Although the use of silicon-on-insulator (SOI) wafers facilitates the sample preparation as the box oxide can be used as a stopping layer [2], more expensive SOI substrates are needed and special process runs have to be executed to deposit the thin-film stacks to be analyzed onto SOI wafers. An approach for backside XPS analysis is desired which uses standard silicon wafers. This paper presents an optimized procedure for backside XPS sample preparation using common silicon wafers and demonstrates its potential to gain insight into the interactions of high-k layers during annealing steps.

EXPERIMENTAL DETAILS

The lapping and polishing steps are carried out with a PM5 precision lapping and polishing machine from Logitech using a PP5D precision polishing jig (see figure 1). The lapping is done using a cast iron lapping plate with a 3 μm aluminum oxide suspension at 40 rpm and a load of 800 g. The polishing is done using a polyurethane plate with a 20 nm silica suspension at 70 rpm and a load of 1500 g. The used glue is Epoxy Bond 110 and the cleaning solution for silica particles is a Micro Organic Soap; both are from Allied High Tech Products.

Figure 1. Setup for plan-parallel lapping (a) and polishing (b) used for backside preparation.

For correcting for any deviations from parallelism during lapping, the sample is mounted on a special 3° facet sample holder and is lapped in such a way that a three-sided pyramid is obtained (sample is rotated by 120° for each facet lapping). The thickness of the lapped sample is monitored using the built-in depth gauge of the PP5D jig for the first few hundreds of micrometers. For measuring the sample thickness in the range of about 60 to 3 μm during polishing, the width w of the sides of the truncated pyramid is measured and the thickness t is then calculated from the equation:

$$t = w * \tan(3°) \tag{1}.$$

The sample thickness below 3 μm is obtained by the observation of interference fringes in a stereo microscope using a red filter of 590 nm wavelength. The thickness t is obtained using the equation:

$$t = a * \lambda / (2 * n) \tag{2}$$

where a is the number of fringes being observed in the microscope, λ is the filter wavelength and n is the optical index. Hence, there is a vertical Si depth of 74 nm for each observed fringe.

The XPS analysis is performed on a Theta300 system from Thermo Instruments in a parallel angle resolved mode. 16 different angles are recorded simultaneously between 20 and

80 degrees, as measured from the normal of the sample. Standard procedures are used for calculating the layer thickness and concentration of the different layers. Monochromatized Al Kα radiation is used, leading to an overall resolution (full-width at half-maximum of the Si 2p peak) of ~0.9 eV.

SAMPLE PREPARATION

The sample is first cleaned in isopropyl alcohol (IPA). The sample support called puck is flattened by grinding using SiC paper. This step also improves the adhesion of the glue to the puck. The puck is then soap cleaned, rinsed in deionized (DI) water and IPA before drying. The sample is glued up side down onto the puck, the two parts are pressed together and the glue is oven hardened at 130°C (figure 2a).

The sample is then mounted to the 3° facet sample holder and a three-sided pyramid is lapped (figure 2b and 2c). The lapping of one side is finished when the puck surface appears. The resulting pyramidal tip should be about in the center of the puck. Note that the lapping time for each pyramid side reduces from about 1h for the first side to about 5 min for the third side. The lapping is then continued with a standard sample holder (no tilting) until a sample thickness of about 60 μm is achieved.

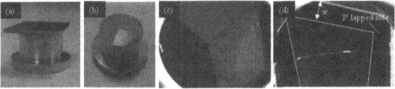

Figure 2. Sample at various stages of lapping and polishing procedure: sample is glued to the puck (a); a three-sided pyramid is formed (b); zoom-in of lapped pyramid showing rough lapped surface (c); a flat triangle forms during polishing, a scratch in the center is observed on this particular sample (d).

The sample is then polished whereby the pyramid is becoming increasingly truncated and a flat triangle forms (see figure 2d). The width w of the three pyramid sides is regularly measured and the remaining thickness t is calculated using equation 1. The difference in value for w for the three pyramid sides is an indicator for the flatness of the sample. Corrections can be made by adjusting the angle on the jig using three micrometer screws.

When reaching a thickness of about 3 μm, interference fringes start to appear and are used for thickness measurement. A circular interference pattern forms in case of a well-polished sample. The number of fringes is counted and the thickness is calculated using equation 2. The larger the distance between two fringes, the more flat the area is. The polishing is continued until a final thickness of about 300 nm. The puck is then soap cleaned and is rinsed in DI water. Figure 3 shows two examples of finished samples, one with an inner area of 56 μm in diameter (figure 3a) and another one with an inner area of about 2 mm in diameter (figure 3b). Note that such results are only obtained with a plan-parallel polishing tool and an optimized procedure.

Figure 3. Use of interference fringes for endpoint detection: four fringes indicating that a Si thickness of 296 nm is reached (a); a large flat area of ~2 mm in diameter is obtained on this particular sample (b).

The remaining silicon is etched away in 15% KOH at 40°C and is then rinsed in DI water. An only 1 nm thick SiO_2 layer is used as an etch stop and therefore extensive overetching should be avoided. The backside sample is cleaned in 4% HCl to remove Fe particles from the surface which originate from the KOH etching [4]. The sample is finally rinsed in DI water and is dried. Note that a total layer thickness of about 15-20 nm is everything what remains from the originally 775 μm thick Si wafer substrate.

Figure 4 shows atomic force microscopy (AFM) images taken on a finished backside sample after the final cleaning step. It can be seen that some silica polishing particles remain on the surface even after soap cleaning. A root-mean-square (RMS) roughness of 5 nm is obtained for the 30x30 μm^2 scan and an RMS value of 0.3 nm is obtained for the 2x2 μm^2 scan. The vertical line structure in the background of the AFM image 4a is attributed to the rough puck surface caused by the lapping step. Hence, the smoothness of the puck surface should be further improved in order to obtain low RMS values of 0.3 nm also for larger areas. Furthermore, more effective cleaning steps must be developed to remove the silica particles completely from the surface (a well-known problem in the area of chemical-mechanical polishing).

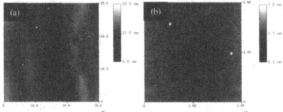

Figure 4. AFM measurement of polished sample: 30x30 μm^2 scan (a); 2x2 μm^2 scan (b).

XPS ANALYSIS

Figure 5 is an example for a non-annealed high-k metal gate stack consisting of 1 nm SiO_2 + 2 nm HfO_2 + 0.7 nm La_2O_3 + 10 nm TaN. The optical microscope image shows an area in the center where all silicon seems to have been removed with some remaining silicon in an area on the top and on the bottom (figure 5e). XPS mapping is carried out before the actual XPS measurement to select the best area on the sample. The different XPS maps in figure 5 for Hf,

Ta, SiO$_2$ and Si confirm that all silicon has been removed in the center region, the SiO$_2$ layer has not been completely removed during KOH etching, and that the HfO$_2$ and TaN layer are still present. Therefore, the actual XPS measurement was carried out in an area close to the center region. Figure 5f shows the relative depth plot of the angle-resolved XPS (ARXPS) measurement which provides a qualitative depth distribution of the different species. This feature is excellent to check on the ordering of the different layers but it does not provide quantitative information. As can be seen, all layers are shown at the correct location in the stack which also validates the correctness of the measured profile. The obtained layer thicknesses from the constructed profile have been added to figure 5f and they agree very well with the nominal values. The strongly reduced SiO$_2$ thickness is most likely related due to KOH etching (KOH attacks SiO$_2$ at a rate of about 6 nm/h at the used conditions) but it cannot be excluded completely that the SiO$_2$ layer is not completely closed (e.g. presence of pinholes). Figure 5g shows the reconstructed elemental profile where the presence of carbon at the surface is linked to surface contamination. There seems to be free oxygen present in the stack which indicates that the SiO$_2$ layer might not be completely closed and therefore H$_2$O adsorption occurs. Note also that the position of the La$_2$O$_3$ layer is shown correctly in the relative depth plot (figure 5f) but it looks intermixed with HfO$_2$ in the elemental profile (figure 5g). This could be linked to H$_2$O adsorption as well or be caused by the increased roughness of the puck (figure 4a). The shape of the Si, Hf, Ta and La signal looks normal as it is expected from an as-deposited stack. It should be mentioned that such XPS measurements are relatively time intensive; the XPS mapping takes 13 h and the actual XPS measurement takes 12 h.

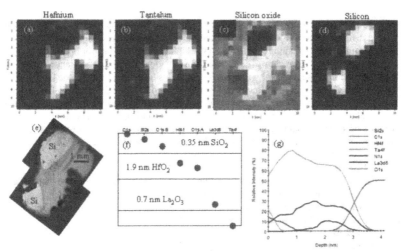

Figure 5. Backside XPS analysis: XPS mapping of Hf (a), Ta (b), SiO$_2$ (c), and Si (d) signal; optical image of sample (e); relative depth plot (f); reconstructed elemental XPS profile (g).

Figure 6 is an example for a high-k metal gate stack of 1 nm SiO$_2$ + 2 nm HfO$_2$ + 0.9 nm Al$_2$O$_3$ + 10 nm TiN. Here, the influence of a high-temperature annealing step on the intermixing of HfO$_2$ and Al$_2$O$_3$ is studied. Figure 6a shows the relative depth plot of the non-annealed sample which confirms that the layer structure has been preserved by the sample preparation. Figure 6b presents the Ti 2p spectra from the sample before and after annealing. The main structure is attributed to TiN. After annealing, a slight increase of the intensity at ~458 eV binding energy is observed which indicates the formation of an oxidized layer, likely from the oxidation of the interface between Al_2O_3 and TiN. This has been confirmed by X-ray reflectometry (XRR) analysis of the same samples.

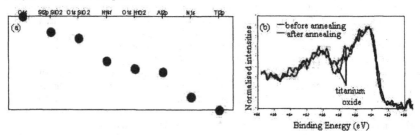

Figure 6. Backside XPS analysis: relative depth plot of non-annealed sample (a); XPS profile comparison of non-annealed and annealed sample indicates presence of an oxidized layer (b).

CONCLUSIONS

Suitable samples for backside XPS analysis can be prepared from common Si substrates by the developed method. This overcomes the need for expensive SOI wafers. The sample preparation involves a combination of lapping, polishing and etching steps. The key for obtaining high-quality backside samples are plan-parallel polishing and the use of the interference-fringe method for endpoint detection. Initial XPS mapping helps to locate a suitable area to perform the XPS measurement. First results demonstrate the power of backside XPS analysis for studying high-k metal gate stacks and illustrate the influence of annealing steps on the layer composition. The removal of the silica polishing particles should be further improved. The developed sample preparation procedure should also enable the analysis of ultra-thin films by other analysis techniques and the same area could be studied by different techniques (e.g. AFM, XPS, SIMS).

ACKNOWLEDGMENTS
The authors would like to thank Fred Stevie and Roberto Garcia from the North Carolina State University for the excellent polishing training. Ole Sumpf and Glen van Vugt from Struers are acknowledged for supporting the first polishing experiments. Alain Moussa from IMEC is acknowledged for the AFM measurements.

1. S.J. Klepeis, *Mat. Res. Soc. Symp. Proc.* **115**, 179 (1988).
2. K.L. Yeo, A.T.S. Wee, R. Liu, C.M. Ng, and A. See, *Surf. Interface Anal.* **33**, 373 (2002).
3. C. Gu, A. Pivovarov, R. Garcia, F. Stevie, D. Griffis, J. Moran, L. Kulig, and J. F. Richards, *J. Vac. Sci. Technol. B* **22**, 350 (2004).
4. C. Bergenstof Nielsen, C. Christensen, C. Pedersen, and E. V. Thomsen, *J. Electrochem. Soc.* **151**, G338 (2004).

Mater. Res. Soc. Symp. Proc. Vol. 1184 © 2009 Materials Research Society   1184-HH08-08

Composition Quantification of Microelectronics Multilayer Thin Films by EDX: Toward Small Scale Analysis

T. Conard[1], K. Arstila[1], T. Hantschel[1], A. Franquet[1], W. Vandervorst[1,2], E. Vecchio[1], S. Burgess[3] and F. Bauer[4]

[1] IMEC, Kapeldreef 75, 3001 Leuven, Belgium
[2] Instituut voor Kern- en Stralingsfysica, K. U. Leuven, Celestijnenlaan 200D B-3001 Leuven, Belgium
[3] Oxford Instruments NanoAnalysis, Halifax Road, High Wycombe, HP12 3SE, UK
[4] NanoAnalysis, Oxford Instruments GmbH Analytical, Otto-von Guericke-Ring 10, 65205 Wiesbaden

ABSTRACT

In order to continuously improve the performances of microelectronics devices through scaling, SiO_2 is being replaced by high-k materials as gate dielectric; metal gates are replacing poly-Si. This leads to increasingly more complex stacks. For future generations, the replacement of Si as a substrate by Ge and/or III/V material is also considered. This also increases the demand on the metrology tools as a thorough characterization, including composition and thickness is thus needed. Many different techniques exist for composition analysis. They usually require however large area for the analysis, complex instrumentation and can be time consuming. EDS (Energy Dispersive Spectroscopy) when coupled to Scanning Electron Microscopy (SEM) has the potential to allow fast analysis on small scale areas.

In this work, we evaluate the possibilities of EDS for thin film analysis based on an intercomparison of composition analysis with different techniques. We show that using proper modeling, high quality quantitative composition and thickness of multilayers can be achieved.

INTRODUCTION

The most recent technology generation for complementary metal-oxide semiconductor applications is characterized by the implementation of alternative materials such as high-k dielectrics and metal gate electrodes in the transistor gate stack. Although a viable solution exists and has been brought to the level of full-scale production for the 45 nm technology, research efforts for high-k materials have not reduced significantly. For future technology generations, the integration of high-k with Ge and III-V materials is envisaged. A major challenge here is to obtain a proper electrical passivation of the interface, which is complicated by the extreme sensitivity of these materials to process conditions and material interactions already at relatively low thermal budgets. Next to these developments for logic devices, novel dielectrics are researched for several other applications to accommodate the requirements as specified by the International Technology Roadmap for Semiconductors. Memory applications, both volatile dynamic random access memory and nonvolatile flash are in direct need for alternative high-k materials and/or device concepts to the ones currently employed. As the requirements for all these applications are significantly different, the chance of finding a common dielectric stack is negligible. This implies that material screening and application of specific gate stack engineering become again more important.

The integration of high dielectric constant high-k materials, such as HfO_2 or $HfSiO_4$, as gate oxide layers into metal-oxide-semiconductor (MOS) devices also requires replacing the conventional polycrystalline Si electrode with an appropriate metal electrode. This becomes necessary because of dopant diffusion across ultrathin gate dielectrics and especially because carrier depletion of the Si electrode increases the EOT of the gate stack to an unacceptable degree. This problem can be alleviated by using a metal electrode because of its high carrier density. However, the band alignment at a metal-high-k dielectric interface cannot be simply tuned by adjusting the electrode doping level as in the case of polycrystalline Si and proper material development is needed

In this article, we will cover three different applications. First, we will investigate the determination of growth mode of ALD (Atomic Layer Deposition) HfO_2 layers on GaAs, where the quantitative determination of the material deposited in the first cycles of the growth is critical for achieving high quality oxide layers. Second, we will investigate the composition of Ta-based metal layers in function of process fabrication. Finally, we will apply quantitative analysis on multi layer systems.

RESULTS AND DISCUSSION

Experimental

The EDS experiments were performed with a Oxford Instruments INCA Energy 350 Microanalysis system including a 30 mm^2 INCA PentaFET-x3 Si(Li) detector. The resolution of this system is better than 133~eV at Mn $K\alpha$ for count rates up to 4000. The measurements in this work were typically performed with 5 mm working distance, scanning area of 100x100 μm^2, beam intensity adjusted to give 2000 cps count rate, and a data collection time of 10 min.

Measuring ThinFilms using EDS

Energy Dispersive X-ray Spectroscopy (EDS) microanalysis in conjunction with Scanning Electron Microscope (SEM) imaging is the routine tool used in micro-scale materials research where spatially resolved compositional characterisation is required. Electron-sample interaction, and the measurement of resultant X-rays can be used in the measurement of nano-metre scale thin films [1]. The basis of the technique is to measure X-ray line intensities for each element and adjust the unknown parameters in a simulation model until the predicted intensities match those measured.to determine mass thickness and /or composition as relevant for each sample. The model is constructed based on user input information of the elements present in the layers and substrate

The analyzed volume and the emitted fraction of the x-rays depend on three factors: energy of the electron beam, element specific energy of x-ray studied, local atomic weight and density. All the varieties of experimental conditions require a realistic distribution in depth of primary generated intensities ($\varphi (\rho Z)$). The k-ratio values (Ix / Io) are determined by the quantitative XPP element analysis [2] . These k-ratios are used together with the depth excitation function $\varphi (\rho Z)$ for calculating the thickness in mass thickness $\mu g/cm^3$ and the weight concentration wt% of elements. Giving the correct density of layer it's possible to calculate thickness and composition of multilayer systems.

Reduction or optimization in accelerating voltage will change the depth distribution of X-ray emission by large factors improving ability to resolve information from surface layers [3]. Accuracy and sensitivity of analysis of stratified samples will depend on the choice of operating conditions and measured lines because of the change in emitted x-ray spectrum by the layer structure itself. In the case of stratified specimens with thin layers (<100nm) the interaction volume from which characteristic x-rays are generated will not in general be confined to one layer but will extend into the substrate and over several layers in the case of multilayer structures.

Growth mode determination

ALD is a technique allowing the deposition of very thin layers in a highly conformal way. It is based on the subsequent self-limiting chemical reaction separated in time. In the case of the growth of HfO_2 in this work, $HfCl_4$ and H_2O are the two reactants leading to the formation of HfO_2. $HfCl_4$ reacts with –OH groups present at the surface of the sample, liberating HCl to form Hf-O bounds and leaving Hf-Cl$_3$ groups at the surface. The H_2O in turn can react with the Hf-Cl$_3$ groups to liberate HCl and leave OH groups at the surface, allowing cycling these reactions. It has been shown however that the start of the growth is highly related to the surface preparation, leading possibly to either growth inhibition or growth enhancement [4]. For instance, the growth on H-terminated Si surface is strongly inhibited, leading to 3D-growth instead of a 2D growth. The growth on oxidized Ge surfaces on the other hand leads to the growth of several layers in one cycle. It is thus also important to carefully determine the amount of deposited atoms at the start of the growth.

The experiment was performed on GaAs wafers covered by their native oxide. The XPS (X-ray Photoelectron Spectroscopy) As$3d$ and Ga$3d$ spectra for GaAs substrates without clean (native oxide) are shown in Figure 1. The native oxide layer of the GaAs substrate consists of As_2O_3, As_2O_5, and Ga_2O_3 and is ~0.7 nm thick. The XPS results show from the reduction of the intensity of the As peak at ~45 eV and the Ga peak at ~21 eV that some reduction of the $GaAsO_x$ layer occurs upon growth of the HfO_2 layer. Rutherford backscattering Spectrometry (RBS) Backscattering has also shown that this system leads first to enhance growth per cycle at the start, followed by static growth cycle afterwards [4].

Figure 2.a presents the EDS spectra from the Hf M-lines for varying number of cycles in the growth on GaAs. All spectra were normalized to the Ga L-line intensity. It is clearly observed that even for the lowest achievable coverage, clear spectra allowing quantification are obtained.

An initial sample model consisting of a layer on of HfO_x on GaAs was tested . The results are presented in Figure 2.b (red squares). When comparing the layer thickness obtained with RBS coverage (that can be converted in thickness through the assumption of the density), one sees that all data seem overestimated and mostly that the intercept for no coverage is at 7.5E14 at/cm^2. In the first version of our model, comparing HfO_x thickness and RBS coverage assumes implicitly that all films have a constant density and composition. One major difference between the RBS quantification and the EDS one is that RBS only takes into account the Hf intensities, while in EDS, the whole spectra, including the oxygen lines is used to retrieve the layer thickness. As observed in the XPS spectra, a thin $GaAsO_x$ layer is present at the surface of the wafer before deposition and the presence of the oxygen from the interfacial layer has thus an influence on the calculated HfO$_x$ layer thickness.

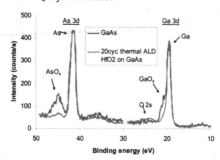

Figure 1: As3d and Ga3d spectra from the starting GaAs surface (black) and after deposition of 20 cycles of HfO$_2$ by a thermal ALD process (red).

In the first model, a fixed density has to be used for the HfO$_x$ layer. However, for thin layers, the product density times thickness is fully independent of the density assumed. In order to improve in the quantification, it is thus necessary to use the HfO$_x$ composition determined by the ThinFilm software, together with the assumed density and the layer thickness. The improvement is visible in Figure 2.b and an intercept of 2E14 at/cm^2 is obtained (blue rhombs).

Using the knowledge of the system from the XPS measurements, a more sophisticated model can be developed, with a two-layer structure composed of GaAsO$_3$ with varying thickness and an HfO$_2$ layer with varying thickness. The results of this model are presented in Figure 2.b (green triangles). In this case, a linear regression shows an intercept at -3E14 at/cm^2. However, as the number of unknowns to be calculated has increased, the statistical errors estimated by the ThinFilm software are larger in the case of the bi-layer model than with the single layer model. yet In cases of both models, the correlation between RBS and EDS is excellent with a correlation coefficient of 0.9986.

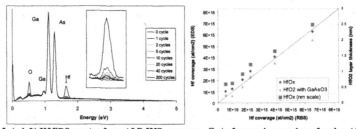

Figure 2: (a-left) Hf EDS spectra from ALD HfO$_2$ grown on GaAs for varying number of cycles. All spectra were normalized on the maximum Ga intensity at 1.1 keV. (b-right) Correlation of Hf coverage as measured by EDS using different models (see text) and RBS measured coverage. The line is a guide for the eyes giving the 1:1 correlation

The detection limit of Hf in this experiment is estimated at ~5E14 at/cm^2 showing clearly the possibilities of EDS for this kind of experiment. In a similar case, where Al$_2$O$_3$ was deposited on GaAs, the observed detection limit was ~1E15 at/cm^2.

Layer bulk composition and thickness

In order to achieve correct band alignment of the gate in transistors, metals with the correct effective work function have to be developed. As the workfunction is dependent on the film composition, correct composition control has to be achieved. One of the metal-system candidates is Ta-based material, mostly carbo-nitride, in which oxygen may be incorporated. The exact composition of a material may depend both on the process method and on the process parameters. Comparing materials produced by different methods requires also precise material composition analysis. Due to the presence of light elements (C, O, N), RBS is not the most suited method for quantitative analysis of thin films. Elastic recoil detection (ERD) is a good alternative but is not commonly available. It has been shown that a good correlation between bulk composition of TaCNO layers of ~50 nm measured by ERD and XPS and Auger in sputtering mode can be achieved [5] However, using sputter profiling in Auger or XPS is also very time-consuming.

One additional difficulty arises from the fact that metal layers are often deposited on SiO_2 layers in order to be able to determine their work function. Determining the presence of oxygen in the bulk of the metal film is thus difficult with techniques that are not intrinsically depth resolved. However, through the use of EDS and multi-layer modeling, significant improvement can be achieved. The system considered here was constituted of ~50 nm TaCNO layers deposited on 10nm SiO_2 layers. Through aging, the top surface of the layer was also oxidized.

A model of the samples was thus developed in a three layer structure. The bottom layer with 10 nm SiO_2; a central layer of unknown composition and thickness of TaCNO and a top layer of 2 nm Ta_2O_5 representing the native oxide formation. The thickness and composition in the lower and top layers were fixed parameters in order to achieve a mathematically solvable problem. The results of the quantification of the TaCNO layer are presented in Figure 3. A reasonably good correlation is observed, at least as good as the one obtained with XPS and Auger. This clearly shows the applicability of the method for film composition analysis.

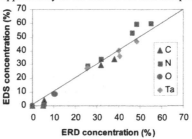

Figure 3: Correlation of atomic concentration measured in the bulk of TaCNO layers by ERD and the composition determined by EDS through a multi layer modeling.

The thin film measurement technique fundamentally calculates mass thickness (the number of deposited atoms per unit area). Therefore, if thickness is known by other techniques, the simulation software can be very useful to measure the density by simply multiplying the density used in the simulation by the ratio of the measured and known thickness. Due to its good sensitivity, and good reproducibility (standard deviation measured on the Ta concentration measured at several point on one given sample ~1%), the technique can be used in quality control to measure the thickness of deposited layer when the density is known.

Multi-layer bulk composition and thickness

One of the major problems in the use of Ge for high performance transistor applications is the development of an adequate surface passivation. Several schemes have been studied for this purpose, including sulfur passivation, stable GeO_2 layer growth, etc.. One possibility is also to combine layers of several materials to benefit from each of their most important characteristics. For instance, limited interaction of Al_2O_3 is observed with Ge, while ZrO_2 is a material with higher dielectric constant that may allow reaching a lower EOT. Combining a thin Al_2O_3 layer as interfacial layer with a ZrO_2 layer is thus also an attractive scheme.

A set of 3 samples of sulfur-passivated Ge surfaces were prepared with a double stack Al_2O_3/ZrO_2 with varying thickness of the two layers. The total layer thickness (nominal) was 2 nm. Figure 4 presents the correlation of the thickness measured by XPS and EDS. A good correlation is obtained between the two techniques, despite the fact that the layers were significantly less than 2 nm in thickness. One can also observe that the total stack thickness is even better correlated than the independent ones. Given the fact that the XPS quantification is done based on theoretical sensitivity factors and from an electron mean free path estimated by an empirical formula, and that the EDS quantification is based on bulk densities, the correlation can be considered as very good. It should however be mentioned that for such thin layers, EDS is not able to define the layer order. We indeed observed that using a model with ZrO_2 as top of bottom layer has no influence on the quantified results.

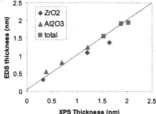

Figure 4: Correlation of thickness measured by XPS and EDS for a Ge/Al$_2$O$_3$/ZrO$_2$ stack

CONCLUSIONS

In this work, we evaluated the possibilities of EDS for thin film analysis based on an intercomparison of composition analysis with different techniques. We show that using proper modeling high quality quantitative composition and thickness of multilayers can be achieved.

REFERENCES

1 J.L. Pouchou, Analytica Chimica Acta, (1993) 283 p81

2 J.L. Pouchou, F. Pichoir and D. Boivin, Proc. ICEM 12, Seattle 1990; Microbeam Analysis, San Francisco Press, (1990), 120; ONERA Report TP 1990-109.

3 I. Barkshire et al., Microchimica Acta 132 (2002) 113-128

4 Delabie, A.; Brunco, D.; Conard, T.; Favia, P.; Bender, H.; Franquet, A.; Sioncke, S.; Vandervorst, W.; Van Elshocht, S.; Heyns, M.; Meuris, M.; Kim, E.; McIntyre, P.; Saraswat, K.; LeBeau, J.; Cagnon, J.; Stemmer, S. and Tsai, W. Journal of the Electrochemical Society. Vol. 155: (12) H937-H944; 2008.

5 IMEC-Matsushita internal data, to be published

Mater. Res. Soc. Symp. Proc. Vol. 1184 © 2009 Materials Research Society　　　1184-HH05-05

Improving the High-Temperature Tensile Strength of SiTiC/TiAl Micro Composites by Interface Engineering

Takakazu Suzuki
National Institute of Advanced Industrial Science and Technology (AIST), 1-1 Higashi, Tsukuba, Ibaraki 305-8564, JAPAN

ABSTRACT

Silicon carbide (SiC) fiber-reinforced titanium aluminide (TiAl) micro composites have been studied using transmission electron microscopy (TEM). C coating can produce a remarkable improvement in mechanical tensile strength up to 1200 K. Coating by a refractory metal, such as W, can impart strength up to 1400 K. C coating by chemical vapor deposition is particularly effective, providing not only a weak interface but also a diffusion barrier. A design concept to increase the tensile strength of SiTiC/TiAl composites is suggested.

INTRODUCTION

Titanium aluminides tend to have the very attractive properties of low density and an excellent oxidation resistance. However, they suffer from low creep strength, arising in most cases from an inadequate ductility and toughness. Moreover, titanium aluminides show a distinctive behavior, increasing mechanical strength with increasing temperature up to 1000 K.

Intermetallic matrix composites, such as titanium aluminides, exhibit improved mechanical properties when reinforced with heat resistive fibers, such as silicon carbide (SiC) fibers. SiC fiber-reinforced TiAl composites [1-9] have attracted attention due to their application in aerospace systems, such as advanced turbine engines and hypersonic vehicles, where strength and stiffness at high temperatures are critical. The interfacial layer of SiC/TiAl composites is expected to be a major contributor to mechanical properties. A variety of interface coatings have been applied to SiC fibers using different interface coating systems. The results indicate the potential for high-performance fibers and composites.

Much research on the interface properties of SiC/TiAl composites has been conducted, but most of the research dealing with transmission electron microscopy (TEM) of the interface of fiber reinforced composites was conducted from a longitudinal-sectional view. Research with a perfect cross-sectional view to obtain more direct, reliable interfacial information has been lacking due to the difficulty of preparing thin TEM specimens without damaging the interface.

Here the author presents a process for producing perfect cross-sectional TEM specimens of SiTiC/TiAl composites. The resulting interfacial characterization leads to a proposal for the manufacture of SiTiC/TiAl composites.

EXPERIMENTAL DETAILS

A polymer-derived pylorized SiTiC fiber (Tyranno fiber; 60 % Si- 2 % Ti- 15 % C- 13 % O) ~ 13 µm in diameter was used. The fiber contained Ti to improve thermal stability [10] and its tensile strength was ~ 3000 MPa. The fiber was coated with ~ 500 nm of C by chemical vapor deposition (CVD) and/or ~ 500 nm of W by sputtering. Finally, a layer of TiAl ~ 1500 nm thick was deposited by magnetron sputtering at a power of 200 W [2, 9].

The ultimate tensile strength of the SiTiC/TiAl composites was measured with a Minebea TCM 50 (NMB) tester. The testing samples of composite, which had been annealed for 2 hrs in Argon (Ar) at a specified temperature between room temperature (RT) and 1400 K, were fixed onto a paper holder with an epoxy adhesive. The holder was cut into two parts before testing. A gauge length of 20 mm and a crosshead speed of 8 mm/s were applied. Tensile strength was calculated from the breaking load and the filament's diameter observed using scanning electron microscopy (SEM). Tensile strengths of more than 30 samples for each annealing temperature in Ar were measured, and statistically treated using two-parameter Weibull analysis [11].

TEM specimens were prepared by sandwiching, and 3 mm disks were obtained by ultrasonic drilling and mechanical polishing to ~ 100 µm. GATAN 600 C ion milling was used for forming thin films for TEM observation. Disks were further ground by a dimpling machine to ~ 10 µm, and Argon-ion-etching was used to get an electron-beam-transparent foil. A Hitachi 9000 HR II microscope with an acceleration voltage of 300 kV was used.

RESULTS and DISCUSSIONS

Changes in SiTiC/TiAl micro composite tensile strength

The relative tensile strength of an SiTiC fiber/TiAl micro composite is defined as the strength measured after TiAl coating and annealing, divided by the original SiTiC/TiAl tensile strength without interface modification or annealing. Relative tensile strength decreases with increasing annealing temperature, the exact behavior depending on the presence (or absence) and type of coating (Fig. 1). However, all relative tensile strengths of SiTiC/M/TiAl micro composites with a modification layer (M) considerably exceed the relative strength of unmodified SiTiC/TiAl at every annealing temperature.

The highest relative strength (~ 1.5) is shown for the case of M = C, followed by the case of M = C/W. This seems to support the weak interface concept [12, 13]. The strength of SiTiC/C/TiAl micro composites after annealing at more than 1100 K was drastically reduced compared to the relative strengths for the cases of M=W and M=C/W. For M=W and M=C/W, relative strengths of SiTiC/M/TiAl composites were significantly higher (~ 0.7) than the relative strengths of the other composites after annealing at more than 1200 K.

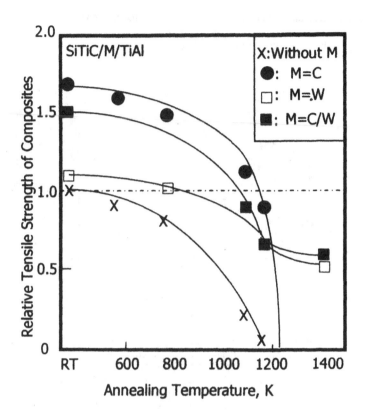

Fig. 1 Change in tensile strength of SiTiC/M/TiAl for various modification layers (M).

This means that, at these temperatures, the tungsten has been able to protect the fiber from the interface chemical reaction because of its extremely high oxidation resistance. However, the weak point was the low relative strength (~ 1.0) of SiTiC/W/TiAl in a temperature range from RT to ~ 1200 K. This seems to arise from brittleness. Overall, using M = C/W in SiTiC/M/TiAl seems to be a suitable modification for higher temperature applications.

Interface characterization of SiTiC/M/TiAl micro composites by TEM

To ascertain the causes for the observed changes in SiTiC/M/TiAl relative tensile strength, the author studied the interface of the SiTiC fiber and the TiAl matrix using TEM.

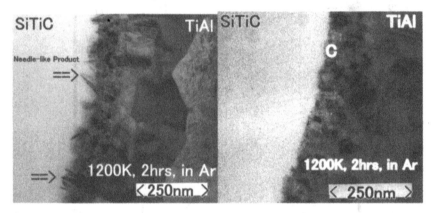

Fig. 2 (a) A TEM cross section of SiTiC/TiAl and (b) a TEM cross section of SiTiC/C/TiAl.

At the SiTiC/TiAl interface many needle-like structures were found sticking into the fiber (Fig. 2(a)). These seem to be the diffusion products of a reaction. The SiTiC fiber itself seems to deteriorate due to the presence of these innumerable needle-like structures sticking into the fiber [12, 14]. In contrast, there were few needle-like structures at either the TiAl/C or the SiTiC/C interfaces, as shown in Fig. 2 (b). This is due to the high resistance of the C layer to diffusion and the inhibition of the chemical reaction between the SiTiC fiber and the TiAl matrix.

Lattice spacing assignment of the needle-like structure

Fig. 3(a) shows a needle-like structure at the interface between the fiber and TiAl in a TEM image of an unmodified SiTiC/TiAl micro composite annealed at 1200 K for 2 hrs in an Ar atmosphere. Since a clear lattice image was observable in some areas, the author has tried to calculate the lattice plane spacing, based on the Joint Committee on Powder Diffraction Standards (JCPDS). The lattice plane spacing was estimated to be ~ 0.24 nm (see Fig. 3(b)), very close to d= 2.390 Å which is the spacing of the most intensive (111) plane of titanium aluminum carbide (Ti$_3$AlC)$_5$C [15, 16, 17].

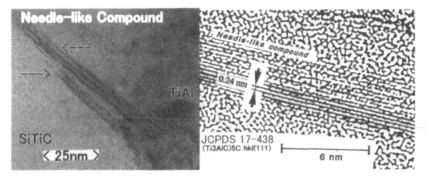

Fig. 3 (a) TEM of needle-like structure and (b) the corresponding high resolution phase-contrast image and calculation.

Reactivity of SiTiC fibers with TiAl and micro composite degradation

There are two possible fracture mechanisms which can be applied to the SiTiC/TiAl micro composite system: fiber cross section reduction [18] and notching at stress concentrations [19]. SiTiC fibers should be relatively unreactive due to their highly amorphous structure [20].

However, strength should decrease significantly if any locally weak sections develop that are highly reactive due to the presence of large defects, such as the needle-like structures that were observed sticking into the SiTiC fibers at the interface of SiTiC/TiAl.

CONCLUSIONS

Based on these results, the following innovative design concept for enhanced-performance composite materials is proposed. When TiAl is reinforced with SiTiC fibers, modification with a carbon layer is essential. This structure provides not merely a weak interface condition, which favors higher performance TiAl composites, but also a diffusion barrier layer that may inhibit the chemical reaction between the fiber and the TiAl matrix. To achieve higher temperature stability or increased oxidation resistance an extra coating of a refractory metal should be considered.

This guideline is based on the results of a SiTiC fiber/TiAl micro composite study. Further study of bulk composites is required before practical use of this method can be unambiguously recommended.

ACKNOWLEDGMENTS

We thank Ms. Y. Tanimoto, Dr. H. Umehara, and Dr. X-L. Guo for help and discussions.

REFERENCES

1. N. S. Stoloff, and D.E. Alman, Mat. Res. Soc. Bull. **12**, 47 (1990).
2. T. Suzuki, H. Umehara, and X.L. Guo, in Proc. of ICCM-12 edited by T. Massard, A. Vautrin (Paris, France 1999) 462.
3. K. Shimoda, J-S Park, T. Hinoki, and A. Kohyama, Compos. Sci. Technol. **68**, 98 (2008).
4. C. Balazsi, K. Sedlackova, Z. Czigany, Compos. Sci. Technol. **68**, 1596 (2008).
5. S. Mall and J. L. Ryba, Compos. Sci. Technol. **68**, 274 (2008).
6. C. Even, C. Arvieu, and J. M. Quenisset, Compos. Sci. Technol. **68**, 1273 (2008).
7. X-L. Guo, T. Suzuki, H.Umehara, in Proc. of ICCM/7 edited by D. Hui (Denver, Co. 2000) 837.
8. D. Koch, K. Tushtev, and G. Grathwohl, Compos. Sci.Technol. **68**, 1165 (2008).
9. H. Umehara, T. Suzuki, and R. Hayashi, J. Japan Inst. for Metals **8**, 1138 (1994).
10. G. Chollon, R. Pailler, and R. Naslain, J. Mater. Sci. **32**, 327 (1997).
11. W.J. Weibull. Appl. Mech., **18**, 293 (1951).
12. T. Suzuki, H. Umehara, and R. Hayashi, J. Mater. Res. **9**, 1028 (1994).
13. D. B.Gundel, S.G Warrier, and D.B. Miracle, Compos. Sci. Technol. **59**, 1087 (1999).
14. T. Suzuki, X-L Guo, H. Umehara, and S. Terauchi, J. of Mater. Sci. Lett. **18**, 1799 (1999).
15. R Jenkins, JCPDS International Centre for Diffraction Data (Swarthmore PA 1990) 17-438.
16. W. Jeitschko, H. Nowotny, and F. Benesovsky, Monatsheft fur Chemie **95**, 319 (1964).
17. W. Jeitschko, H. Nowotny, and F. Benesovsky, Kohlenstoffhaltige ternare verbindrungen (H-phase), Monatsheft fur Chemie **94**, 672 (1963).
18. G. Blankenburges, J. Aust. Inst. Metals **14**, 236 (1969).
19. P.W. Jackson, Metal Eng. Q **9**, 22 (1969).
20. T. Yamamura, T. Ishikawa, M. Shibuya, and K. Okamura, J. Mater. Sci. **23**, 25 (1988).

Mater. Res. Soc. Symp. Proc. Vol. 1184 © 2009 Materials Research Society   1184-HH08-10

Quantitative Characterization of the Interface Between a V_2O_3 Layer and Cu_3Au (001) by Cs Corrected HREM

H. A. Calderon[1], H. Niehus[2,3], B. Freitag[4], D. Wall[4], F. Stavale[3] and C. A. Achete[3]   ·

[1]Depto. Ciencia de Materiales, ESFM-IPN, 07738 Mexico D.F.
[2]Inst. für Physik, Humboldt Universität zu Berlin, Newtonstraße 15, Berlin 12489, Germany
[3]Div. Metrologia de Materiais (DIMAT), Inmetro, Xerem, RJ, Brazil and Prog. Engenharia Metalúrgica e de Materiais (PEMM), Univ. Fed. Rio de Janeiro, RJ, Brazil
[4]FEI Company, Achtsweg Noord 5, Eindhoven, The Netherlands

ABSTRACT

Vanadium oxides are materials of interest due to their electronic, magnetic and catalytic properties. In the case of V_2O_3 and Cu_3Au, the interfacial bonding is rather difficult to describe since the two component materials have strongly different electronic structures. Thus a local investigation of the interface becomes important. In this investigation, the incoherent interface between a V_2O_3 (0001, corundum structure) layer and a Cu_3Au (001, $L1_2$ structure) substrate is characterized with the help of image corrected high resolution electron microscopy (HRTEM) and focal series reconstruction in order to investigated both the true position of atoms and the nature of the atomic species. Semi-quantitative results can be shown for the chemical composition of columns and strains at one side of the interface.

INTRODUCTION

Interfaces between oxides and metals play an important role in the field of electronic packaging, gas sensors and combustion engines. Vanadium oxides have interesting properties attracting considerable and wide interest (e.g., [1-3]). There is normally a need to find the mechanisms responsible for such properties and thus the characteristics of regions near and at the interface itself require a full and detailed investigation. These interfaces join two materials with completely different electronic structures making the interfacial bonding difficult to be explained by a particular chemical bonding state. Additionally there is normally a lattice misfit that when relatively small, a coherent pseudomorphic interface is produced. In such cases, severe stress and strain fields can develop and give rise to a two-dimensional network of self organized misfit dislocations structures. Increasing the lattice mismatch even further, generates an incoherent interface with a complete lack of lattice continuity across it and where an understanding in terms of a specific pair bond model is meaningless. This type of interfaces has been investigated by modeling with different techniques [4, 5]. However, there is a considerable need of reliable experimental data for comparison and further advance in the understanding of these interfaces.

The method to grow different thin oxide layers on a $Cu_3Au(001)$-Oxygen Surface (CAOS [6]) has been developed recently. The surface properties of the sesquioxide have been described in detail theoretically [7] and experimentally [8-9]. There is a lattice mismatch between V_2O_3 ((0001), corundum structure) and Cu_3Au ((001), $L1_2$ structure) that gives rise to an incoherent interface (see Fig. 1). Scanning tunneling microscopy (STM) measurements show large areas of a two dimensional $V_2O_3(0001)$ layer exposing absolutely no stress induced features [9]; a similar behavior has been reported for ultra thin NbOx layers [6]. The only crystallographic relation other than the configuration of the (001)Cu$_3$Au surface plane interface (see Fig. 1). Scanning tunneling microscopy measurements show large areas of a two dimensional V$_2$O$_3$ layer exposing absolutely no stress induced features [9]; a similar behavior has been reported for ultra thin NbO$_x$ layers [6]. The only crystallographic relation other than the configuration of the (001)Cu$_3$Au surface plane parallel to (0001)V$_2$O$_3$ appears in the azimuthal alignment of the oxide with respect to the substrate, being determined by an alignment of the <1100>V$_2$O$_3$(0001) and the <1120>V$_2$O$_3$(0001) with the corresponding <110>Cu$_3$Au (001) directions (Fig. 1b).

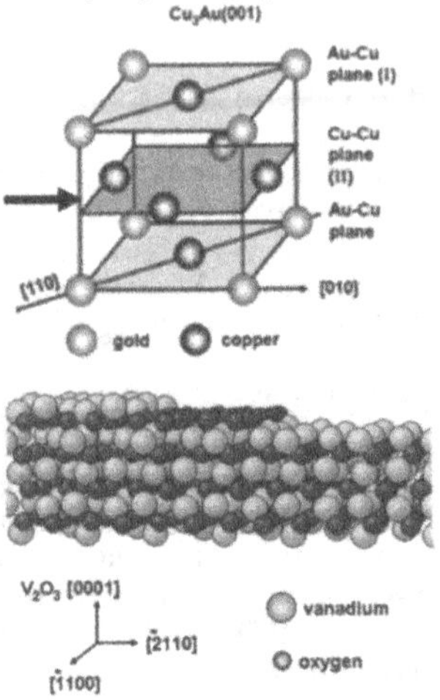

In this work, the characteristics of the incoherent Cu$_3$Au-V$_2$O$_3$ interface are investigated further on the basis of high resolution aberration corrected electron microscopy. This is a relatively new development of the technique that allows direct interpretation of images. Quantitative measurements made under such conditions are reliable and help understand the interface properties.

Fig. 1. Crystallographic structures (a) L$_{12}$ of Cu$_3$Au, (b) Corundum structure of V$_2$O$_3$.

EXPERIMENTAL PROCEDURE

The metal-oxide pair to characterize in this work is vanadium sesquioxide grown on a Cu$_3$Au(001) substrate by the CAOS method [6]. In this case it is possible to grow a two dimensional (2D) multilayer of V$_2$O$_3$(0001) on top of the Cu$_3$Au(001) substrate. The substrate has been prepared in ultra high vacuum to produce a flat Cu$_3$Au(001)-oxygen surface [10, 11] by oxygen implantation and subsequent annealing at 650 K. In a next step about 1.5 nm vanadium is evaporated at 300 K sample temperature followed by 6 min annealing at 850 K in 2 x 10^{-7} mbar oxygen ambient. The V$_2$O$_3$(0001) layer has been identified in situ by the typical LEED 'ring'-type superstructure (due to 90° rotated domains). Finally a vanadium protection layer of about 10 nm is evaporated on the entire sample surface. The samples for electron microscopy have been prepared in a dual beam (Scanning Electron and Focused Ion Beam) apparatus (Nova 600 FEI ®). The thin slices used for transmission have been cut normal to the Cu$_3$Au(001) surface along a <110> direction (see the arrowed rectangle in Fig. 1a). These samples are cleaned by glow discharge and then transferred to a Titan microscope with Cs image correction for the HRTEM investigations. The images were taken at 300 KV and with an objective lens Cs of 2 μm. The exit wave reconstruction has been performed with focal series of 20 images taken from a starting zero defocus in steps of -2 nm. The software package TrueImage (FEI ®) is used for this procedure. The required instrumental parameters had been previously determined e.g., focus spread (3.7 nm), sampling rate (0.01608 nm/pixel), semiconvergence angle (0.1 mrad) and the MTF (modulation transfer function) at Nyquist frequency (0.3) which is related to resolution. Experimentally, the exit wave reconstruction requires approximately 10 min to be completed and provides valuable information regarding position and nature of the atomic columns but it can only be applied to sufficiently thin sample regions with no contrast variation. Additionally the focus position needs to be known with good approximation through the use of Thon rings. The whole procedure is described in [12]. It can be considered as a safer route to microscopic analysis in comparison to single direct images, especially for cases where there is little limitation due to sample damage. In the present work, thin areas close to the sample edge have been used in an effort to satisfy the requirements of the reconstruction procedure without a formal determination of thickness. However TrueImage goes through a rather elaborated testing of the experimental data by performing image simulation that must match to the experimental images. Such testing ensures (if passed) that the sample is sufficiently thin. All phase and amplitude images shown here passed all the quality tests of TrueImage.

RESULTS

The oxide film quality and thickness is shown in the HRTEM survey image presented in Fig. 2. Here the interface, the Cu_3Au substrate, the vanadium oxide and the polycrystalline protective V layer can be seen. The left side (darker area) represents the plane projection perpendicular to the original $Cu_3Au(001)$ surface along a $<110>$ direction. The sharp interface is characterized by a line of bright dots (see also Fig. 3) followed to the right side by the crystalline V_2O_3 film in the middle section which is protected by the polycrystalline vanadium capping . Two different structural orientations (domains) can be found in the V_2O_3 layer as shown in Fig. 3. Domain 1 is associated with a zig-zag structure (Fig. 3a)

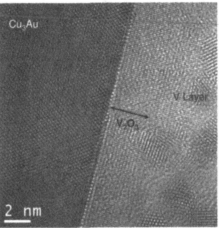

Fig. 2. Overview of V_2O_3 layer on top of a Cu_3Au substrate viewed along [110]. The protective V capping layer is also shown.

while Domain 2 has closer interatomic distances and a rectangular pattern (Fig. 3b). These two domains represent only different views of the same $V_2O_3(0001)$ crystal structure, the view perpendicular to a $<1100>$ direction (domain (1)), and a $<1120>$ direction (domain (2)), respectively. Analysis of these images allows to propose a model for the interface. This is also shown on the basis of domains 1 and 2 in this work (Fig. 3e).

Qualitatively, Fig. 3 can be used to extract valuable information about the incoherent interface between Cu_3Au and V_2O_3. For instance, Fig. 3a shows a brighter contrast for the V_2O_3 on the right side and only the row at the interface with the Cu_3Au substrate remains very bright. The border or interface line is marked with a two headed arrow in Fig. 3a, the brighter spots still retain the Cu_3Au ordering. A local thickness variation is a possibility for interpretation, however the blobs with higher intensity are present in all parts of the interface and only there (see Fig. 2), making such a possibility rather weak. On the other hand, it is significant that the intensity of these blobs compares in magnitude with the intensity of the Vanadium atoms located on the V_2O_3 side of the image. Measurements of intensities are given in Figs. 3c,d, the location of such line scans is marked by arrows in Fig. 3a. Thus apparently V atoms play an important role in the formation of the interface. The V atoms at the interface follow approximately the L_{12} structure and thus their position differs from the corresponding one in the corundum structure i.e., they do not belong to the vanadium oxide part. As a result, it is proposed (following [13] and the model in Fig. 3e) that a new monolayer of vanadium is introduced at the interface exposing features expected by the onset of an alloy formation for V_3Au (001). Fig. 3a also shows a brief identification of the atomic columns by using similar symbols as in the model. Nevertheless, this conclusion requires further support from electron microscopy e.g. the presence of similar intensity features in phase and amplitude images obtained after reconstruction since interpretation of Cs corrected microscopy can be further enhanced by such a procedure.

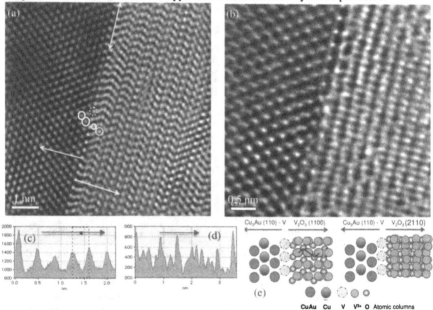

Fig. 3. Domains and proposed Cu_3Au (001)-V_2O_3 interface configuration as seen along [011]. (a), (b) Domains 1 and 2 of V_2O_3 with the location of intensity scan profiles. (c) Intensity scan corresponding to Cu_3Au and (d) oxide sides, the arrow indicates the scan direction away from the interface. (e) Interface models with atomic columns in projection for both domains.

The Vanadium plane (brighter spots at the interface) is displaced with respect to the Cu_3Au structure introducing a small structural distortion that can be measured with precision. Quantitative measurements in nanostructures can obviously be done on direct images. But the intensity in such images is always a mixture of the amplitude and the phase of the exit wave and the determination of the atomic positions can be obscured by this fact. A more reliable procedure requires the determination of the exit electron wave or image reconstruction. The full technique has been described elsewhere [12]. Accordingly, a phase and an amplitude images can be created from the reconstruction procedure, containing accurate information regarding the position of the atomic columns and its particular intensity is related to the chemical composition of the atomic column. Additionally the use of such a technique allows correction of several aberrations and renders a higher precision to the measurement. Figure 4 shows an example of the obtained results after image reconstruction. A section of the complete amplitude and phase images is shown in Figs. 4a,b. The magnitude of the intensities found in direct images (e.g. Fig. 3) are reproduced i.e., low intensities at the interfacial atomic row in the case of amplitude image and a high intensity for the corresponding phase image. This supports the proposed model for the interface. In this work, the amplitude image is used for quantitative evaluation of the lattice parameters. This is done by directly finding the positions of the blobs i.e. locating the intensity minima and then measuring the corresponding distances between them. Fig. 4c shows an example, the location of the intensity minima is given by white crosses. The rotation between the images in Figs. 4a and 4c is necessary for a better performance of the evaluation software (DARIP TotalResolution®). The software produces a file with the positions of the blobs (X and Y) and its intensity. The use of a data sheet (e.g. Excel) allows determination of distances between atomic columns along selected directions. The results for two different directions (indicated by double headed arrows in Fig. 4c) are given in Figure 5, only results for the Cu_3Au side are analyzed here.

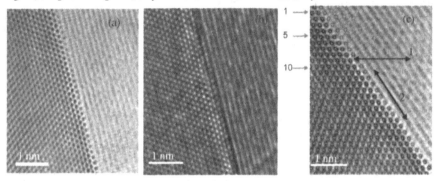

Figure 4. (a) Amplitude and (b) phase image obtained from exit wave reconstruction in the interfacial region and (c) location (white crosses) of the intensity minima for the quantitative evaluation. The two-headed arrows indicate the measuring directions (1 and 2) and the numbers (left side) give the different averaged rows.

All interplanar distances on the Cu_3Au side are evaluated in pixel units in the amplitude image. After measurement, they are then averaged according to the specific row that they occupy. For clarity, some rows are identified by numbers in Fig. 4c. Two measuring directions are selected, direction 1 takes columns along a horizontal line in Fig. 4c and direction 2 produces averaged interplanar distances in lines parallel to the interface. The averaged values per line are then plotted in Fig. 5a after converting them into linear distances by using the lattice parameter of Cu_3Au (0.374 nm) as reference. They clearly depend on all lattice distances measured along a particular row. Thus for direction 1, some rows produce averages which are statistically weighted to values of lattice distances measured near the interface or far way from it. For instance line 1 produces an average interplanar distance representative of a region near the interface and line 25 produces an average value weighted to distances far away from the interface. Therefore the observed variation in lattice distance (see diamonds in Fig. 5) can be used as an indicative of lattice straining at the interface. The measurements along direction 2 (rows parallel to the interface) have the advantage of considering a higher number of measurements so that the corresponding average values have a higher statistical significance, the results are shown also in

Fig 5 by means of circles. As for the error bars, they have been computed from the standard deviation in the measurements made away from the interface (Cu_3Au lattice) where a small to nonexistent distortion can be expected. As can be seen in Fig. 5 and in both measuring directions, the averaged interplanar distance clearly varies a function of position with respect to the interface. Thus the presence of the V layer at the interface introduces a considerable strain on the Cu_3Au side. The information is presented also in a different manner i.e., a lattice mismatch here called strain is plotted against row number (right side Y axis). The relationship used for the strain parameter is $(d_x-d_{away})/d_x$, where d_x represents a given interplanar averaged distance and d_{away} is a distance far away from the interface. A strain or lattice mismatch of around 8 % is found close to the interface along the measuring direction on the Cu_3Au side for both measuring directions.

CONCLUDING REMARKS

The development of an interface in this system apparently implies complicated mechanisms. The formation of a V row of atoms with an L_{12} structure at the boundary region surely has to do with minimization of energy i.e. a driving force towards equilibrium is reducing the interfacial energy. Under such conditions it is likely that the measured strain (around 8 %) represents also a compromise to satisfy a minimum energy criterion.

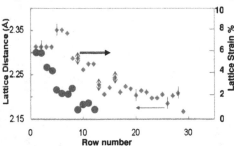

Figure 5. Evaluation of interplanar distances on the Cu_3Au side up to the interface. Average lattice distances and lattice strain as a function of row number of atomic columns. Measurement directions are indicated in Fig. 4c. The horizontal arrows indicate the reference (Y axis) for the error bars

ACKNOWLEDGMENTS.

The support of INMETRO; IPN (COFAA, SIP) and CONACYT (58133) is acknowledged.

REFERENCES

1. V.E. Henrich, P.A. Cox, *The surface science of metal oxides*, U. Press, Cambridge, 1994.
2. K. Hermann, M. Witko, in *The chemical physics of solid surfaces: Oxide surfaces*, vol. 9, Ed. Woodruff D.P., Elsevier, Amsterdam, 2001.
3. G.A. Somorjai, Introduction to Surface Catalysis, Wiley, New York, 1994.
4. R.-P. Blum *et al.*, Phys. Rev. Let. **99**, 226103 (2007).
5. A.E. Mattsson, D.R. Jennison, Surf. Sci. **520**, L611 (2002).
6. J. Middeke, R.-P. Blum, M. Hafemeister, H. Niehus, Surf. Sci. **587**, 327 (2005).
7. I. Czekaj , K. Hermann, M. Witko, Surf. Sci. **545,** 85 (2003).
8. A.-C. Dupuis et al., Surf. Sci. **539**, 99 (2003).
9. H. Niehus, R.-P. Blum, D. Ahlbehrendt, Phys. Status Solidi (a) **187**, 151 (2001).
10. H. Niehus in *The chemical physics of solid surfaces: Oxide surfaces*, vol. 9, Ed. Woodruff D.P., Elsevier, Amsterdam, 2001.
11. H. Niehus, C. Achete, Surf.Sci. **289**, 19 (1993).
12. W. M. J. Coene *et al.*, , Ultramicroscopy 64, 109 (1996).
13. H. Niehus *et al.*, Surface Science **602**, L59 (2008).

Printed in the United States
By Bookmasters